Renewable Energy Systems

Renewable Energy Systems sets out what is possible for anyone interested in using renewables at home or for small businesses. Explaining in clear detail the best options for implementing renewable and energy efficiency technologies, it will help lay to rest some common misconceptions.

Fully illustrated, it provides an expert overview of precisely which sustainable energy technologies are appropriate for widespread domestic and small business application. The book is separated into sections for each technology, and detailed case studies, installation diagrams and colour photographs show the possibilities for the average household, workshop and office, as well as for small- to medium-sized commercial or public buildings.

Technologies covered include:

- wood energy
- wind power
- solar photovoltaics
- solar thermal
- passive solar
- geothermal
- air-to-air heat pumps
- hydro-based energy systems

plus the all-important subject of energy efficiency.

Ideal for anyone that wishes to become involved in the industry, this complete overview of the technologies in a clearly accessible format will be invaluable for students, building engineers, architects, planners, householders and small business owners.

Dilwyn Jenkins was Executive Director at the Powys Energy Agency between 2000 and 2004 and presently works from Wales as a consultant and writer on environmental issues, mainly related to energy, rainforest conservation, community development and cultural travel. Much of his work has focused on projects in Peru and Brazil as well as in Wales, Italy, Scandinavia and Austria. He is the author of *Wood Pellet Heating Systems* (Earthscan, 2010).

Earthscan Expert Series
Series editor Frank Jackson

Solar:
Grid-Connected Solar Electric Systems
Geoff Stapleton and Susan Neill

Solar Domestic Water Heating
Chris Laughton

Solar Technology
David Thorpe

Stand-Alone Solar Electric Systems
Mark Hankins

Home Refurbishment:
Sustainable Home Refurbishment
David Thorpe

Wood Heating:
Wood Pellet Heating Systems
Dilwyn Jenkins

Renewable Energy Systems

The Earthscan Expert Guide
to Renewable Energy Technologies
for Home and Business

Dilwyn Jenkins

Series Editor:
Frank Jackson

First edition published 2013
by Routledge
2 Park Square, Milton Park, Abingdon, Oxon, OX14 4RN

Simultaneously published in the USA and Canada
by Routledge
711 Third Avenue, New York, NY 10017

Routledge is an imprint of the Taylor & Francis Group, an informa business

© 2013 Dilwyn Jenkins

The right of Dilwyn Jenkins to be identified as author of this work has been asserted by him in accordance with sections 77 and 78 of the Copyright, Designs and Patents Act 1988.

All rights reserved. No part of this book may be reprinted or reproduced or utilised in any form or by any electronic, mechanical, or other means, now known or hereafter invented, including photocopying and recording, or in any information storage or retrieval system, without permission in writing from the publishers.

Trademark notice: Product or corporate names may be trademarks or registered trademarks, and are used only for identification and explanation without intent to infringe.

While the author believes that the information and guidance given in this work are correct, all parties must rely upon their own skill and judgement when making use of them – it is not meant to be a replacement for manufacturer's instructions and legal technical codes. The author does not assume any liability for any loss or damage caused by any error or omission in the work. Any and all such liability is disclaimed.

Every effort has been made to contact and acknowledge copyright owners. If any material has been included without permission, the publishers offer their apologies. The publishers would be pleased to have any errors or omissions brought to their attention so that corrections may be published at a later printing.

British Library Cataloguing in Publication Data
A catalogue record for this book is available from the British Library

Library of Congress Cataloging-in-Publication Data
Jenkins, Dilwyn.
Renewable energy systems : the Earthscan expert guide to renewable energy technologies for home and business / Dilwyn Jenkins. – 1st ed.
  p. cm. – (Expert series)
1. Renewable energy sources. I. Title.
TJ808.J46 2012
621.042–dc23
2011050794

ISBN: 978-1-84971-369-6 (hbk)
ISBN: 978-0-203-11726-2 (ebk)

Typeset in Sabon by Keystroke, Station Road, Codsall, Wolverhampton.

Contents

Illustrations vii

Preface and Acknowledgements xvii

1 Introduction to Renewable Energy Technologies for the Home and Business   1

2 The Importance of Energy Efficiency   13

3 Wood Log and Wood Pellet Heating   51

4 Solar Thermal   71

5 Heat Pumps   91

6 Photovoltaics   111

7 Wind Energy   131

8 Hydropower   163

9 Case Studies   193

10 Annex 1: Finance   215

11 Annex 2: Resources   227

Index   237

Illustrations

Figures

1.1   2 kW micro hydro intake showing small reservoir, intake and penstock (pipe taking water to turbine lower down the stream's course)   3
1.2   Small-scale wind turbine providing power to household in hillside location   5
1.3   Small house with solar absorbers (solar water heating panels) and photovoltaic (solar electric) modules on same south-facing roof   6
1.4   Wood pellet stove with back boiler heating room, domestic radiators and providing household hot-water needs   7
1.5   Heat pump and hot water cylinder providing heating to standard-sized household   8
2.1a–b   Average home energy use in the USA and UK. In the UK the largest use is space heating alone, while in the USA, with its higher temperatures in many places, heating and cooling are the biggest energy users   16
2.2a–b   The two main types of natural ventilation: (a) stack (or convective) ventilation and (b) cross ventilation. Natural ventilation can create greater comfort and savings, particularly where cooling is required   17
2.3   Simple structural devices like roof overhangs can be designed to minimise solar gain in summer and maximise it in winter   18
2.4   Workings of an air source heat pump operating in cooling mode   19
2.5   Optimising natural light inside building and helping to heat the space by maximising solar gain via extensive surface areas of glass (in walls, windows and skylights) at the Centre for Alternative Technology, Wales, UK. The wall to the left provides thermal mass heat storage, being made of rammed earth   20
2.6a   Testing the air tightness of a home using a special fan called a blower door can help to ensure that air sealing work is effective. Often, energy efficiency incentive programs such as the DOE/EPA Energy Star program require a blower door test, usually performed in less than an hour, to confirm the tightness of the house   21
2.6b   A blower door in situ. Door blower tests use a powerful fan to pull air out of a building to create exaggerated but detectable drafts   21
2.7a–b   External and internal photograph of well draught-sealed letterbox with brush in the middle and sprung flaps on both sides   22
2.8   This diagram illustrates a number of simple and standard techniques for keeping an indoor space warm: insulation (in ceiling, wall, floor and loft), e-coated double glazing and natural solar gain through windows   23
2.9a   In a cool climate, the majority of heat lost from an uninsulated, free-standing house is through the walls, followed by the roof. However, the roof has a much smaller surface area than the walls indicating that more is lost per square foot from the roof. In terms of installing insulation, this is usually where the biggest and easiest savings can be made   24

2.9b	Air infiltrates into and out of your home through every hole and crack. About one-third of the air infiltrates through openings in your ceilings, walls and doors	24
2.10	Extended surface areas of glazing and Trombe type walls can be used for optimising solar gain and an element of heat storage	25
2.11	An attached conservatory or sun room can be used effectively to reduce heat loss and provide a naturally sun heated space	26
2.12a	A pack of sheep's wool insulation – a natural fibre insulation for the healthy home	28
2.12b	A pack of hemp insulation	28
2.12c	A sample block of rigid foam insulation with reflective backing for extra thermal benefits	28
2.12d	A sample block of thick cork insulation	28
2.13	Sheep's wool insulation bats with the source of their raw material	30
2.14a–b	Sheep's wool insulation being fitted into timber-framed building wall-panels at the Centre for Alternative Technology, Wales, UK	30
2.15	Photograph showing tan-coloured internal wall which has been insulated and re-plastered simultaneously with a lime and hemp mix	30
2.16	Photograph of dry-lined wall filled with flax insulation rolls	31
2.17	External insulation (white) being applied to existing wall, Berlin, Germany	32
2.18	If it's not possible or appropriate to put insulation under the concrete floor, then an insulation layer can be laid on top, either between floor joists for a wooden top-surface, or under a second layer of concrete	34
2.19	When it comes to suspended wooden floors, the boards need to be lifted and the insulation laid between the joists and, ideally, filled to the full depth of the available space. Vapour barriers should not be fitted in case of liquid spillage	35
2.20	Double-glazed window unit with gas-filled space between the two glass panes	35
2.21a–d	External storm windows can protect the window sashes and original glass, as well as reducing energy loss	36
2.22	Using a Watt meter in a home or office can help identify which appliances are consuming lots of electricity and at what times, a good starting point for deciding how to invest in more efficient equipment or measures	38
2.23a	Compact fluorescent light bulb (CFLs). These typically use around 75% less energy than a conventional incandescent bulb to provide the same level of lighting (in many countries incandescent bulbs are no longer allowed to be sold)	39
2.23b	Light emitting diodes (better known as LEDs) are even more efficient than CFLs. They also last longer and do not give off heat	39
3.1a	A modern wood stove with heat resistant glass door	52
3.1b	Wood-pellet stove with back-boiler	52
3.1c	Cutaway showing combustion process in wood-pellet boiler	52
3.1d	A Stanley solid-fuel cooking range with back boiler (large fuel box designed to 17 kW with wood logs)	52
3.1e	A 540 kW KOB biomass boiler	52
3.2a–b	A well-constructed log store with slatted sides for airflow and a decent roof to keep the rain off	53

3.2c	Not quite as well constructed a log store, it has a roof but not enough airflow since the side walls are filled	53
3.2d	A rough-and-ready wood store, it has a roof which might blow off in high winds causing safety risks and the side is too open allowing any heavy rain to dampen the logs	53
3.3	Masonry Heater, by Brian Klipfel of Fireworks Masonry	56
3.4	Wood pellets are clean and dust-free to handle	59
3.5	Wood pellet boiler and footprint	60
3.6	Wood pellet automated fuel-feed mechanisms: (a) screw auger; and (b) vacuum operated	61
3.7	Fully automated wood-pellet boilers make them a strong contender for home, office or workshop situations	63
3.8	Fuel Price Comparison UK, late 2010	64
3.9	Fuel price comparison USA, 2011	64
3.10	Wood stoves need regular de-ashing, but pellet boilers tend to keep this to a minimum, sometimes requiring it as little as twice a year. This ash chamber has a compacting device	65
3.11a–b	(a) The Ludlow woodstove and (b) The Stretton inset stove from AGA, both of which are smoke control exempt	67
4.1a	Daily annual average solar radiation on south-facing surfaces tilted at the angle of latitude in kWh/m^2/day, US	71
4.1b	Yearly sum of solar radiation on south-facing optimally inclined surfaces in kWh/m^2, Europe	72
4.2	A simple form of solar shower, from a company that specialises in camping equipment	72
4.3	Schematic of a solar pool-heating system showing all the main components	73
4.4	Schematic of a solar thermal system with twin coils in the hot water tank. Heat from the solar panel is pumped into the bottom of the tank when the sensor on the panel detects that it has reached the required temperature. Another sensor in the hot water tank determines whether the heat needs to be topped up by the second coil, fed from the domestic boiler. Even during the winter the solar panel will provide some pre-heating of the water in the tank – even a few degrees will give the boiler less work to do.	75
4.5	Clip-fin collector, self-assembled. These can be purchased in kit form.	76
4.6	Typical flat-plate collector. This Sol 25 Plus Collector has a surface area of 2.7 m^2 (29.06 ft^2).	77
4.7a	Diagram showing how a flat-plate collector works	78
4.7b	Diagram showing how a clip-fin collector works	78
4.8a–c	(a) Display at the Centre for Alternative Technology, UK, showing on the indirect and direct evacuated-tube systems; (b) and (c) Close-ups of evacuated tube collectors	79
4.9	Evacuated tubes contain vacuums to retain more heat. Fins, or a cylindrical surface that maximises the area of collector directly facing the sun, are attached to, or enclose, the fluid-containing tubes. This increases the period over which heat can be collected, both during a day and throughout the year. The surface is given a coating, which absorbs more heat radiation frequencies. Evacuated tubes are more efficient than flat-plate collectors mainly because they do not re-emit as much heat thanks to the vacuum within the tubes	79

4.10a–d	Mirrors can be used inside or outside a tube collector	80
4.11	A Sunflow System with integral storage. Within the casing a reflective mirror concentrates heat onto a stainless steel collector/ cylinder containing 25 gallons of water. The heated water from the collector is drawn from the top of the first collector to the bottom of the second, where it is raised to an even higher temperature	81
4.12a	Cutaway of a twin-coil solar hot-water tank showing a simple layout for stratification, without sensors	82
4.12b	An Okofen twin-coil solar hot-water tank, with a section removed to show the tank's wall-thickness and the two coils within	82
4.13	Schematic of a typical integrated collector storage system with electrical backup	83
4.14	A fully filled, indirect sealed solar thermal system layout. It includes an expansion vessel to contain any expansion of the fluid when it gets hot	84
4.15a–b	Diagram and photo of a thermo-syphon system with integrated storage tank	86
4.16	Solar collectors for space heating on a Passivhaus-standard office building in Neudorfer in Rutzemos, Austria	87
4.17	An evacuated-tube system that provides domestic hot water as well as radiant underfloor heating	88
5.1	Coefficient of Performance (COP) for heat pumps is usually between three and four; in this example we show it as four, where 75% of the energy comes from sun-heated ground source and 25% is electrical input, either from grid or renewable sources	92
5.2	The heat pump cycle uses a combination of ground-source and electrical energy to transform a liquid from low pressure and low temperatures into a high pressure and high temperature vapour	93
5.3	Coefficient of Performance factors – in this example 3 kW of heat is extracted from the Earth and 1 kW of electricity is put in from the mains to achieve an output of 4 kW of heat, offering a coefficient of performance of four	94
5.4	Seasonal efficiencies of heat pump types: ground source heat pumps (GSHP) are more efficient than air source heat pumps (ASHP)	95
5.5	An air source heat pump providing summer cooling. Heat from the air inside the building is absorbed by the refrigerant and released to the outside	97
5.6	Diagram showing the main components of a heat pump system. In this ground source system heat is taken from beneath the surface of the ground and transferred to a household hot-water system. The heat pump itself runs off electricity	98
5.7a–b	Diagram showing two different configurations for a ground source heat pump. The first (a) shows a horizontal system where the pipes are buried in a shallow trench. The second (b) shows a system where the pipes are installed vertically	99
5.8a	GSHP heat exchanger with shallow sealed-loop system. The pipes are laid horizontally in shallow trenches which are then back-filled	99
5.8b	GSHP heat exchanger with 'slinky' shallow coil. The trenches are filled with coiled pipe which allows more pipe to be installed in a smaller area	99
5.8c	Slinky coils laid out on the surface before they are installed in trenches.	99

5.9	Business tapping into underground low grade heat	101
5.10a	Diagram showing main components of a water source heat pump linked to a domestic underfloor heating system and hot-water tank	103
5.10b	Diagram showing the main elements of a water source heat pump. Like all heat pump systems it includes a heat pump, an evaporator and a condenser – the difference here is simply the location and design of the coil loop	103
5.11	Slinky coils being immersed in a pond	104
5.12	Diagram showing a water source heat pump with a direct (open) heat exchanger. It is the water itself that runs into the pump system	104
5.13a	Schematic showing a house with an air source heat pump for space heating. The outdoor heat-exchange coil is housed separately from the building, connected to a wet heating system in the basement by underground pipes (air-to-water system)	105
5.13b	Diagram showing the main components and possible layout of air source heat pump system, connected to a hot-air heating system (air-to-air system)	105
5.13c	Photograph of a domestic air source heat pump	105
6.1	Typical grid-connect PV system serving domestic or business premises with surplus solar power being sold into the grid, and the grid providing power if not enough is being generated by the solar modules to cover building electricity demand	112
6.2	Photovoltaic cell (polycrystalline type)	112
6.3a–6	Diagram showing cross-section of photovoltaic module and photovoltaic module showing how the solar cells are connected in series	113
6.4a	Close-up photograph of Kyocera polycrystalline PV module	114
6.4b	Photograph of polycrystalline PV module fitted into larger array	114
6.4c	Photograph of grid-connected polycrystalline PV module array on roof of business units with single module PV car-park lighting in foreground	114
6.4d	A PV array with a solar water-heating system installed below it	114
6.5	A solar cell at work	116
6.6a	Monocrystalline cell	117
6.6b	Monocrystalline module	117
6.7	Uni-Solar's thin-film flexible solar modules being installed on a roof	118
6.8	In this hybrid (NA-F135) module the tandem structure is made up of an amorphous and a microcrystalline silicon layer	118
6.9	A hybrid system where electricity and hot water are produced simultaneously. This manufacturer produces two versions, one of which is designed to optimise the electrical output while the other optimises hot water output	119
6.10	On-roof mounted PV array in the UK (in this instance the house is grid-linked and sells surplus electricity into the grid, benefitting from preferential feed-in tariffs)	120
6.11	Flat-roof-mounted system: The Energy Roof at Sheffield Gazeley Blade, UK	121
6.12	Ground-mounted PV modules in system owner's garden, Kent, UK, from the *Earthscan Expert Series* book *Grid-connected Solar Electric Systems*	121
6.13a–c	Building integrated PV at commercial offices, Doxford, UK	122
6.14	The components of a grid-connected system. The modules feed power to the home's appliances through the grid inverter, which produces grid-quality	

	AC electricity. Any surplus is sent to the local electricity network through the grid-feed electricity meter. When the home's demand is greater than can be satisfied by the PV array, such as at night-time, power is drawn instead from the local network through the consumption electricity meter. The Utility Company would bill the home for the balance of the amount of power used and supplied, according to the tariffs agreed for the electricity bought and sold	123
6.15	A grid-connected house with solar thermal and PV on the same roof	123
6.16a	PV array, installed in the late 1990s and designed to meet 10% of the building's electricity demand, at Dulas office buildings, Wales, UK	125
6.16b	Recently completed PV grid-connect installation on roof of office building in Stroud, UK	125
6.17	Diagram of an off-grid PV system	126
7.1	Upwind and downwind turbines showing different wind orientations	133
7.2	Main components of a small wind generator	133
7.3	Vertical-axis turbine offsetting some power at a local government building, UK	134
7.4	Diagram of typical grid-connect wind turbine set-up	135
7.5	Swept area and near maximum power outputs of different sized wind turbines	136
7.6	The zone of turbulence (i.e. the action of wind around local obstacles such as trees and buildings)	139
7.7	This wind turbine is not ideally sited – it is frequently hunting the wind direction due to high turbulence because it is sited very close to buildings and right next to a cliff	140
7.8	Windspeed rating scale based on the shape of tree crown and the degree to which twigs, branches and trunk are bent	141
7.9	Photograph of flagging tree (somewhere between IV and V in the scale given in 7.8 – complete flagging towards partial throwing)	142
7.10	Inside a wind-monitoring data logger	143
7.11	Wind-monitoring mast with solar powered data logger unit attached, which can be downloaded remotely via mobile phone connection	144
7.12	Power curve for Proven WT2500 wind generator	146
7.13	Example windspeed distribution at 13 mph average windspeed (1 mph = 0.44704m/s)	148
7.14a	The Air 40 is marketed as a small battery-charging turbine for remote homes, telecom, industry, lighting etc.	149
7.14b	An installation of three turbines in Beijing, China (these are battery-charging wind turbines produced by US company Bergey)	149
7.15	Rutland Windcharger (Marlec) being used in conjunction with a solar photovoltaic panel to power street lighting – the low friction three-phase alternator gives a smooth and silent output and the low windspeed start up means that even light breezes will provide some generation	150
7.16a	Bergey's 10 kW Excel turbine, installed near the Capitol Building in Oklahoma City, US	151
7.16b	The two-bladed Whisper 500 from Southwest Windpower. Its fibre-glass re-enforced design is designed for harsh environments and high windspeeds	151

7.17a–c	Evance Iskra R9000 5.3 kW (formerly called the Iskra) turbines are used in a range of situations, including rural domestic properties, schools, small farms and light industrial sites	152
7.18a–b	Photographs of Proven wind turbines 11 and 35–2 (the latter being down for inspection)	153
7.19a–b	Proven 11 turbine on 9 m tower and different Proven turbines on various towers	153
7.20	Endurance E-3120 50 KW – a three-bladed horizontal axis downwind turbine – is ideal for larger farms, schools, hospitals and commercial/industrial sites	154
7.21	Turbowinds T500–48 500 kW wind turbine, ideal for business parks and grid sales	154
7.22	Large-scale wind turbines don't seem to bother livestock and function well in windy and rural areas	155
7.23	Large-scale turbines need to be located in windy areas, often found in upland or coastal sites	156
7.24	Many of the newest wind farms are located offshore – this creates a new set of engineering issues for developers and even monitoring the wind can be tricky	156
8.1	The solar-driven hydrological cycle	164
8.2	Making a crude (gross) 'head' measurement	166
8.3	Undershot and overshot water-wheel designs	167
8.4	Common hydro system components	168
8.5	Common hydro system configurations	169
8.6	Sketch showing water catchment area on contoured map in relation to intake, settling tank, penstock and powerhouse	170
8.7	3 kW Pelton turbine at the Centre for Alternative Technology, Wales, UK	172
8.8	Voltage, amps and total generation log at house which consumes the hydro electricity	172
8.9	How a Pelton turbine works	173
8.10	Sketches of different turbine types	174
8.11	Common electrical configurations – generating for own use and/or for sale into the grid	175
8.12	Design specifications in cross-section of a twin-Pelton hydro installation	176
8.13	Small hydro intake with Coanda screen and fish ladder on mountain stream	178
8.14	A typical hydrograph for a micro hydro site shows available flow in cubic metres per second (sometimes as litres per second for smaller systems)	179
8.15	Calculating the cross-section of a stream	180
8.16a	The Harris Compact Micro-hydro	181
8.16b	High head Pelton turbine (5.5 kW)	181
8.17	Watercourse and diversion weir from above intake	182
8.18	Intake and diversion weir diagram	183
8.19a	Intake and diversion weir	184
8.19b	Old and new intakes at diversion weir	184
8.19c	Close-up showing movement of water on coanda-effect intake screen	184
8.20a	First section of penstock	185
8.20b	Simple hole valve in penstock	185

xiv ILLUSTRATIONS

8.21	The turbine house at the community hydro project in Talybont-on-Usk, UK. The intake is at the reservoir further up-river. The owners of the reservoir, Welsh Water, are obliged to maintain the flow in the river and some of this 'compensation' flow is channelled through the turbine	185
8.22	Settling tank, penstock and powerhouse diagram	186
8.23a	Penstock arrives at powerhouse	187
8.23b	Pelton wheel, drive-belt and turbine	187
8.24	Powerhouse with water returned to stream	187
8.25a	Hydro inverter controls located in premises being powered to enable remote monitoring	191
8.25b	Battery bank stored in airy wooden cupboard close to inverter controls	191
9.1	A low-energy business unit with own PV array, built in 1996 (UK)	193
9.2a–b	A photograph showing the use of timber in Unit 1, Dyfi Eco Park business unit, and a sketch which illustrates well the design for the apex of the building to allow for solar gain	194
9.3a–b	Photographs of interior upstairs of Unit 1, showing the solar glare from skylight apex and also the cotton sheet used to resolve the problems caused by glare	195
9.4	Photograph of PV array built over a bicycle shed at the front south-facing corner of Unit 1	196
9.5a–b	(a) annual energy consumption and (b) annual CO_2 emissions for Dulas Unit 1 (1998 to 2010). The years with heaviest gas use had colder winters and the increasing trend for more electricity use is do to with greater staff numbers using more and more gizmos that need plugging in or charging (except for a dip in 2006 to 2008 when staff were moving out as the company acquired an additional business unit on the same park).	197
9.6	Photograph of low-energy house with pellet boiler and solar array	198
9.7	Sketch depicting low-energy house with pellet boiler and solar array	198
9.8	Photograph of mixing system and pipes for underfloor heat distribution	199
9.9	Sketch depicting wood pellet and solar water heating working in combination, Case Study 4	201
9.10	Photograph of standard house with pellet boiler and solar array	201
9.11a–c	Photographs of a 14 kW Ecodan air source heat pump (ASHP) installed in a Gloucestershire village, UK	202
9.12a–c	Various perspectives on a barn conversion with horizontal ground source heat pump system, showing (a) the house itself, (b) the external housing for heat pump and (c) the internally located manifolds, Wales, UK	204
9.13a	Completed roof-mounted PV installation at SWEA offices, UK	208
9.13b	Installing hybrid PV module, lifting module onto SWEA office roof	208
9.13c	Fronius inverters used by SWEA to convert PV output into grid-compatible electricity	208
9.14	Chart showing solar power output profile for the first day in March 2010 at SWEA offices, UK	209
9.15	PV arrays can be seen on two roofs at Dyffryn Cottage's domestic installation	210
9.16	Chart showing monthly PV output (generation) against monthly import of grid electricity. There is a marked increase in PV output and decrease in grid electricity used during the summer months	210

9.17	Hockerton Community's 225 kW Vestas wind turbine installed for local investment in a UK parish	212
9.18	Turning on the valve between the penstock piping and the turbine in the power house at Talybont Energy community hydro-turbine	213

Tables

1.1	Summary of renewable resources required for each type of renewable energy	4
1.2	Comparative costs and financial support for implementing the different renewable energy technologies	8
2.1	Natural fibre insulation materials	29
2.2	A rough guide to the type and effectiveness of various insulation materials	32
2.3	Basic heat loss and boiler sizing form for manual calculation	46
2.4	Showing the delivery temperatures required for a range of standard heat distribution systems	47
3.1	Wood heating systems comparison chart	70
5.1	Criteria for selecting a ground source heat pump	102
5.2	Low temperature water-based distribution (e.g. underfloor) is the most compatible and efficient for use with heat pumps	109
7.1	Effects of the wind at different wind speeds	139
7.2	Comparing two small Proven wind turbines: the Proven 11 (5 kW) and the Proven 35–2 (12 kW)	151
7.3	A rough guide to wind-turbine system costs	159
7.4	Logical phases and steps in the process from feasibility assessment to installation of a wind turbine	160
8.1	Critical criteria for deciding whether hydropower is suitable at a specific site	191
10.1	Estimated financial paybacks for each technology (assuming no tax credits, grants, FiTs, or other financial support)	221
10.2	Energy and carbon paybacks by technology	222

Preface and Acknowledgements

Interest in renewable energy is growing and, faced with rising oil and gas prices, rising energy demand and the pressing issue of climate change, is likely to continue far into the future. There is an extensive range of technologies now commercially available: solar photovoltaics, solar thermal, passive solar, wood energy, wind power, geothermal and air-to-air heat pumps, hydro energy systems, as well as energy efficiency measures. The aim of this fully illustrated book is to guide the reader through this maze, enabling him or her not only to have a good overview but also to be able to understand which technologies are appropriate for domestic and small-to-medium sized business application. It is also designed to enable building managers, building engineers, architects, planners, trades people and engineers who are thinking of investing or getting involved installing in these technologies to make the most cost-effective decisions, as well as provide local government planners, other public servants and advice agencies with a clear overview. Schools and colleges will find it an easy-to-read educational tool.

The sections on different renewable energy options provide detailed descriptions of each technology, describing how it works, the main issues to consider, the range of system concepts available and appropriate system types. The circumstances in which the technology makes sense and in which it does not are clearly outlined, alongside how to assess local renewable resources and how to go about procurement and installation. Case studies, installation diagrams and colour photographs show precisely what is possible for the average household or business.

With many thanks for their major inputs to this book to Frank Jackson, Jacinta MacDermott, Brian Horne, Rachel Amstead, Teilo Jenkins, Max Jenkins, Tess Jenkins, Mike Fell and Kat Hollaway. Also, my sincere thanks for their help to: John Cantor (Heat Pumps Ltd), Emma and Matthew Rea, Phil Horton (Dulas Ltd), Dan Hammond (Dulas Ltd), Andrew Rowbottom (Dulas Ltd), Teresa Auciello (Marlec Engineering), Simon Roberts (Centre for Sustainable Energy), Talybont Energy, Andy Rowlands, Paul Whiteman (Information Department, CAT), Pat Borer, Billy Langley, Dan Curtis, Tim Kirby, George Goudsmit (AES Ltd, Scotland), Owen Callender (SWEA), Ecovision Systems, Bruce Cockrean (Ecotricity), Gary Thomas (Ecotricity), Natalie Schofield (Exled Ltd), Scott Merrick (Bergey Windpower Company), Kevin Brooks (PV Solar Installations Ltd) and Jenny Lampard.

1
Introduction to Renewable Energy Technologies for the Home and Business

The world is undergoing a revolutionary resurgence in the use of renewable energy for generating electricity, heat and cooling. This phenomenon can be seen as a result of diminishing fossil-fuel resources, and, particularly in the last decade, as a logical reaction to the reality of climate change, which stems in large part from the use of coal, oil and gas.

Renewable energy has been used in some forms – especially wood heating, passive solar, wind and hydro – for thousands of years in homes, public spaces, work situations and for agricultural processes. Technological development has been rapid since the 1970s providing new opportunities and greater efficiencies for renewable installations. This fact, combined with an increasing range of government and energy utility incentives, has created a highly positive new scenario for the implementation of renewables not just at an industrial scale but also at the domestic and small- to medium-sized business levels.

Many householders, energy managers and decision makers in businesses or public services are turning to renewable energy sources in order to contribute to the mitigation of climate change. Many are also switching to renewables because it makes economic sense. Others have made the move simply because it's such a positive personal or commercial statement. Central and regional governments across the world are taking actions to encourage the uptake of renewables at the domestic- and large-building level.

In 2008, Governor Arnold Schwarzenegger signed an executive order (S-14–08) requiring that retail sellers of electricity should provide 33 per cent of their load with renewable energy by 2020. Furthermore, the California Energy Commission's New Solar Homes Partnership (NSHP) – a $400 million program – offers incentives to encourage solar installations, with high levels of energy efficiency, in the residential new construction market for investor-owned electric utility service areas. The goal of the NSHP is to install 400 MW of capacity by 2016.

Meanwhile, in Chicago, a Clean & Renewable Energy Strategy within the Chicago Climate Action Plan identifies a goal of reducing greenhouse gas (GHG) emissions by 5.33 MMTCO2e by 2020 through the implementation of clean and renewable energy sources. One of the main strands for achieving this is through the promotion of household and institutional renewable power. The UK's Department of Trade and Industry commissioned a report in 2004 which suggested that by 2050, microgeneration – the production of heat (less than 45 kW capacity) and/or electricity (less than 50 kW capacity) from zero or low-carbon source technologies – could provide 30–40 per cent of the UK's electricity needs and help to reduce household carbon emissions by 15 per cent per annum.

Space heating and cooling are the largest consumers of energy in most homes and offices, accounting for almost 45 per cent of total domestic energy use in the US and over 55 per cent in the UK. In business premises the percentages vary but are also significant. With this in mind, household or business decision makers interested in minimising their contribution to global warming should logically look into the technologies which offset the use of fossil fuel in the provision of heating or cooling. Water heating accounts for an average of 14 per cent of total energy use in US households and a massive 26 per cent in UK homes, so renewable energy technologies which can provide all or part of this requirement are also obvious priorities. Cooking, lighting and appliances account for over 40 per cent in the US, so are an obvious priority there for offsetting, where possible with renewable energy sources.

In this book we explain the main renewable energy technologies that are compatible with domestic or business applications in simple terms, but we start with a chapter outlining the need and options for ensuring that a building is as energy efficient as possible. Maximising the energy efficiency of a property which is to be heated or powered by renewables means that the renewable energy system output and costs can be kept to a minimum.

The book's main focus is on grid-connected situations since this is the situation faced by the vast majority of the English-speaking world. We avoid promoting one technology over another because selecting between renewables depends on a range of factors: primarily access to the appropriate renewable resource. Other factors, such as the availability of financial incentives for renewable energy generation or implementation, are secondary but still important criteria to help choose the right technology.

It's rare to find a house or work premises which has equal access to all the main renewable energy resources: good solar radiation, strong and relatively constant winds, an all-year-round stream or river and locally available biomass for example. In terms of heat-pump technologies (ground, water or air source), while most of us have access to air, this is rarely the most carbon- or cost-efficient option and to install a ground sourced heat-pump requires access to suitable land or ponds.

Choosing the right renewable energy for a particular home or business premises should be based principally on a set of criteria such as those outlined below, but every situation is unique so the technologies are explained individually in the main chapters of this book (Chapters 3 to 8). In Chapter 9 we take a look at a range of relevant case studies, covering each of the technologies. Annex 1 is dedicated to the financing of and financial incentives for renewables, while Annex 2 lists a set of resources for finding out more about the renewable technologies discussed.

Environmental Factors

Appropriate Technologies: Available Resources and Associated Social Issues

If you haven't got sufficient access to the right renewable resource, then the technology dependent on that resource automatically excludes itself as an option. All renewable resources are inevitably site-specific and so the first step in selecting the appropriate technology is to get a good estimation of the resource. How to do this is explained in the relevant technology-based chapters. The social factors include issues such as the potential for obtaining planning permission if this is required in the region for certain technology installations (e.g. wind turbines and their towers) and the likelihood or not of upsetting the neighbours. In Table 1.1 we offer a summary of what to look for in terms of the resource for each of the main technologies described.

In terms of the available resources, hydropower is probably the least widely applicable. Very few properties have legitimate access to the right flow of water and gross head (i.e. the vertical difference between a hydro system's intake and the turbine located below in a powerhouse). This is unfortunate in that, given a good hydro-resource, this is probably the preferred renewable technology for generating electricity. Despite having relatively high initial capital-investment costs, hydro systems (like streams and rivers) are typically designed to run 24 hours a day, seven days a week, all year round. You can't say the same for wind or solar resources in most locations.

Figure 1.1 2 kW micro hydro intake showing small reservoir, intake and penstock (pipe taking water to turbine lower down the stream's course)

Source: Dilwyn Jenkins

Table 1.1 Summary of renewable resources required for each type of renewable energy

Renewable energy technology	Access to the essential resource	Site suitability	Other environmental issues
Wood heating	Access to reasonably priced wood logs, wood chips or wood pellets	Chimney or flue system available at the property or possible to install	Local planning regulations allow wood heating systems (e.g. some inner-city areas have smoke or emissions regulations)
Solar thermal	Sufficient solar radiation (compatible with property's thermal requirements/use patterns)	The orientation of property (usually but not exclusively a building's roof) faces the direction of the sun (i.e. roughly south in the Northern Hemisphere and north in the Southern Hemisphere)	No significant shading of solar installation (e.g. from nearby trees or other buildings)
GSHP (geothemal/ ground or water source heat pumps)	Access to land and subsoil or large-enough pond or watercourse	Suitability of subterranean soil type or water resource	Proximity of land or pond to the property
ASHP (air source heat pumps) for heating or cooling	In this case, the essential resource is 'air' within the appropriate temperatures (8°C/ 17°F or above for optimum carbon efficiency)	Suitable external wall is needed (however, almost all properties will have this, except some high rise buildings)	Local planning regulations relating to external wall fixings (e.g. in historic town centres)
Solar photovoltaic	Sufficient solar radiation	The orientation of property (usually but not exclusively a building's roof) faces the direction of the sun	No significant shading of solar installation (e.g. from nearby trees or other buildings)
Wind energy	High windspeeds with enough consistency throughout the year to cover electricity needs (for own use and/or export to the grid)	Availability of and legitimate access to a suitable site for locating wind turbine and its tower (they cannot be located close to obstructions like buildings or trees which can cause turbulence affecting turbine efficiency)	Proximity of site to property or grid connection (the latter, if the system is just designed for grid power sales) and planning permission (if required) for turbine and tower is possible for the location and neighbours are unlikely to be effected adversely (or object)
Hydropower	The available watercourse has sufficient flow and is constant enough to cover electricity needs (for own use and/or export to the grid)	There is legitimate access to the part of the watercourse that offers enough 'gross head' (i.e. the vertical height difference between the hydro system intake and the hydro turbine)	Proximity of site to property or grid connection (the latter, if the system is just designed for grid power sales) and also planning permission (if required)

Not everyone will live or work in a location which has sufficient wind resources; but even where the resource may exist, few individual properties have access to the right spot for locating a wind turbine on a sufficiently tall tower. Firstly, there are problems of access to an area of land where one would be allowed to erect a turbine. Secondly, there are issues of local wind turbulence, which can impede the efficient functioning of a wind turbine. Thirdly, even if the immediate neighbours don't object to the installation, it may be that the local municipality has stringent conditions around planning permits.

Figure 1.2 Small-scale wind turbine providing power to household in hillside location

Source: Marlec Ltd, www.marlec.co.uk

Solar thermal and solar photovoltaic installations are probably the most widely applicable and, for some solar systems, the simplest to install of all the renewable technologies. Most regions have sufficient solar radiation (sunshine) to warrant the use of solar technologies, but not every building is well oriented towards the movement of the sun. Furthermore, even for well-oriented premises, because of shading from nearby buildings or trees not every property has equal access to the solar resource generally available in their region. Solar water heating (SWH) installations are perhaps the most appropriate of all the renewable technologies: with the right roof orientation most properties could benefit and, even without financial support, the payback on SWH is one of the fastest of all the renewable technologies (particularly for sunnier countries in properties that have relatively high hot-water requirements, e.g. swimming pool heating).

Some of the renewable technologies – particularly solar photovoltaics (PV) which provide electricity from the sun – have been suggested to be of little environmental benefit because of the energy embodied in their manufacture. However, in terms of its own environmental impact, a study by the World Energy Council reported that 1 kW of electricity produced by PV would be linked to CO_2 emissions of between 0.01 kg and 0.1 kg. Electricity generated by a gas power station is estimated to be responsible 0.452 kg CO_2 (IEA/OECD, 'CO_2 Emissions from Fuel Combustion', 1999), making gas responsible for at least four times the quantity of emissions as PV. More than that, as a relatively new technology, PV production is improving in efficiency and minimising its impacts all the time.

Like the solar technologies, most regions of the world have some access to wood in the form of log, wood chips or purpose-manufactured wood pellets. However, countries with more forest cover and a traditional forestry based industry (like Finland, Canada and Austria) are in a much better position to supply the necessary raw materials. Indeed, these countries have been involved in pioneering the development of wood combustion technologies. The further an area is from forestry industries the more costly this form of technology is likely to be. Wood pellets are designed, in part, to help overcome the costs associated with transporting woody biomass for heating (and electricity generation). Pellets are drier and denser than logs or wood chips, which makes them cheaper to move

Figure 1.3 Small house with solar absorbers (solar water heating panels) and photovoltaic (solar electric) modules on same south-facing roof

Source: Martin Ashby

Figure 1.4 Wood pellet stove with back boiler heating room, domestic radiators and providing household hot-water needs

Source: Andrew Warren

around. They are also designed for automation as well as minimal human handling in both delivery to a building and with fuel feed mechanisms (from store to boiler).

Geothermal or ground source heat pumps (GSHP) are only applicable for heating premises that have access to suitable land, subsoil or water (usually a large pond, called water source in that case). If a building has access to nearby land or a reasonable-size garden, this technology is certainly an option for the provision of heating. It works best with lower temperature water-based heat distribution systems, like underfloor heating. Air source heat pumps (ASHP) can be used for heating or cooling, but only work with sufficient efficiency where and when the external air temperatures are greater than 8°C (17°F).

Figure 1.5 Heat pump and hot water cylinder providing heating to standard-sized household

Source: John Cantor

Economic Factors

Costs and Financial Incentives for Renewable Energy Technologies

Capital costs for installing the different renewable energy technologies vary greatly. The good news is that in many regions of North America and Europe there are capital grant incentives or tax rebates available for renewable installations. Preferential feed-in tariffs (FiTs) or net metering for renewable electricity generated are also now widely, but not universally, available which means that it's possible to achieve system-payback relatively quickly and for even small- to medium-sized installations to be seen clearly as investments with predictable financial returns.

Table 1.2 Comparative costs and financial support for implementing the different renewable energy technologies

Renewable energy technology	Relative costs	Access to capital (loans or grants)	Other financial incentives (e.g. preferential feed-in tariffs)
Wood heating	Combustion equipment costs range from lower than conventional heating systems to an additional 75% (or even more) for sophisticated installations	Capital grants are often available (generally up to around 30%) in many regions. Loans are usually available for heating systems (depending on credit status)	In Europe most countries have introduced a Renewable Heat Incentive which offers payments based on calculated kWh of heat produced by wood heating systems

Renewable energy technology	Relative costs	Access to capital (loans or grants)	Other financial incentives (e.g. preferential feed-in tariffs)
Solar thermal	One of the lower-cost renewable technologies with good potential for relatively fast payback	Capital grants are often available (generally up to around 30%) in many regions. Loans are usually available for solar water heating systems or solar related building works (depending on credit status)	In Europe most countries have introduced a Renewable Heat Incentive which offers payments based on calculated kWh of heat produced by solar water heating systems
GSHP (geothermal/ ground or water source heat pumps)	Relatively expensive installation costs, but low running costs compared to many space heating systems	Capital grants are often available (generally up to around 30%) in many regions. Loans are usually available for GSHP installations (depending on credit status)	In Europe most countries have introduced a Renewable Heat Incentive which offers payments based on calculated kWh of heat produced by GSHP heating systems
ASHP (air source heat pumps) for heating or cooling	Relative to GSHP, air source heat pumps cost significantly less (but are less efficient at providing heating, though significantly better at cooling)	Capital grants are sometimes available in many regions. Loans available (depending on credit status)	In Europe most countries have introduced a Renewable Heat Incentive which offers payments based on calculated kWh of heat produced by ASHP heating systems
Solar photovoltaic	Although still relatively expensive by kWh/yr of electricity produced, solar photovoltaics (PV) has come down significantly in price during the last ten years or so and are highly competitive in regions where capital support is available	Capital grants are sometimes available in many regions of Europe and North America. Loans available (depending on credit status)	Some parts of North America and Europe operate net metering or preferential feed-in tariffs which make this technology very competitive. Some North American states have counties and utilities which offer tax rebates on solar PV installation costs and feed-in tariffs
Wind energy	Generally less expensive than either hydro or PV by kWh/yr calculations (but this depends on a good available wind resource)	Capital grants are sometimes available in many regions of North America and Europe. Loans available (depending on credit status)	Some parts of North America and Europe operate net metering or preferential feed-in tariffs. Some North American states have counties and utilities which offer tax rebates on wind power installation costs and feed-in tariffs
Hydropower	Generally expensive in terms of capital costs of installation, but with very long life expectancy and generating 24/7 (which means that over the lifetime the kWh/yr cost is relatively low)	Capital grants are sometimes available in many regions of North America and Europe. Loans available (depending on credit status)	Some parts of North America and Europe operate net metering or preferential feed-in tariffs. Some North American utilities offer tax rebates on hydro installation costs

> **Power and Energy Definitions**
>
> **Power** = the rate energy is produced by an energy generating device or consumed by an appliances
> Units of Power = watt (W) or kilowatt (i.e. one thousand watts, abbreviated to 1 kW)
>
> **Energy** = amount of power produced by a generating device over time or consumed over time by an appliance
> Unit of energy = watt-hour (Wh) or kilowatt-hours (i.e. one thousand watt-hours, usually as abbreviated to 1 kW)
>
> **Examples**:
> A 20 W low energy light bulb on for 10 hours = 200 Wh
> 5 × 20 W low energy light bulbs on for 10 hours = 1000 Wh = 1 kW

Billing and Finances

Energy is usually billed by the kWh. Most utility electricity bills, for instance, are based on meter readings or estimates of numbers of units consumed by a property. These units are almost always kWh. The situation is frequently different with heating bills. Fuel oil, for instance, is mostly sold by the litre or gallon. Gas bills are usually calculated in units measured in kWh, shown on your gas meter as volume of the gas used in cubic feet or cubic metres; then gas companies convert this into kWh for billing. The price charged for each unit of energy varies according to pricing plans or tariffs agreed in contracts with your utility provider. In some cases, for instance in some shared accommodation (e.g. departments in the same block or a housing cooperative), which uses a centralised boiler and fuel supply, heat is sold not by how much is consumed, but by floor space or cubic volume of each individual apartment. The larger the apartment, the higher the charge. However, there are systems which will measure how much energy has been consumed by each apartment. This is common practice in many communal buildings that share a biomass heating system in Scandinavia.

Some Advantages Associated with Generating One's Own Energy

Producing one's own energy on site can have the following advantages:

- emissions reduction;
- money saving (payback depends on resource available, technology selected and the going electricity or fuel rates – a cushion against energy price fluctuations);
- selling electricity to the grid can turn a renewable energy system into a decent financial investment: for instance, with net metering and a system that generates more than the property consumes, or, even better, with preferential

feed-in tariffs. More information on net metering and FiTs can be found in the relevant technology chapters and Annex 1;
- security of energy supply and local control over this (except arguably with wood energy, there is no dependence of fuel being supplied by some distant region, country or business);
- renewables are often the ideal solution for providing remote power supplies (e.g. for sailing boats and where it would be expensive to obtain a grid connection – such as mountain top or island communications systems or signal booster installations);
- stand-alone systems powered by renewable energy (sometimes in a hybrid form where more than one energy source is combined to provide the total requirement) avoid utility bills completely and avoid the need for grid connection. This is particularly useful in remote households way beyond grid access; and
- renewables can also be used as a backup in case of grid failure.

Average Home Electricity

To have any chance of making a serious contribution to a property's electricity demand, the average house would need, for example, a wind system of between 1 kW and 6 kW. The actual size of turbine required will depend on levels of energy efficiency implemented, plus the wind and other location factors. In the US properties tend to have higher electricity consumption than those in the UK, but the range of domestic electricity use is quite large – from about 3,500 kWh a year to about 10,000 kWh a year – so this needs to be assessed on a case-by-case basis. A larger building – for instance a public or business premises – is more likely to require a larger turbine, perhaps in the 10 to 30 kW range. Smaller turbines usually generate between 50 W and 500 W, being designed mainly for battery charging in the leisure market – mainly boats and caravans. There is also a range of small turbines suitable for mounting on buildings, some of which are considered appropriate for the urban environment. Solar electricity (photovoltaics or PV) is also of course an option.

Community and Shared Renewables

In theory all renewable electricity sources can be tapped for shared use and/or shared profit. Community-owned wind turbines and hydropower systems have been springing up in many countries over the last ten years. While some are designed to offset a community's energy needs, others are exclusively or partly implemented to sell electricity to the grid. Renewable heating has a longer history of shared generation and use. Community wood heat boilers, for instance, can pump hot water for heating and domestic water needs to entire villages, towns or city suburbs via district heating networks (DHN). Most DHNs are installed underground, with individual connections taking hot water into the heating systems of each property in an insulated ring of piping.

2
The Importance of Energy Efficiency

In most grid-connected situations for homes and small businesses, when considering using a renewable energy source one should always consider energy efficiency measures first. Without optimum efficiency, most renewable installations make little sense in either financial or environmental terms. Whether you consider using renewable energy for electricity or heating, energy efficiency measures will significantly reduce running costs and will almost always result in less capital expenditure on the generation of electricity or heat. The accepted wisdom across the world for householders and workspace managers alike is that energy efficiency measures should always be the first step in reducing money costs, minimising our carbon footprint and reducing our environmental impact on the planet. As well as saving money and the planet, energy efficiency measures can also improve the levels of comfort in homes and offices. Cold, draughty, damp spaces with a tendency to condensation need to be eliminated altogether, with buildings being healthier living and working environments when energy efficiency measures are fitted properly.

In this section, after clarifying what energy efficiency means, we look at energy efficiency measures for the home and office before going on to cover a range of specific energy efficiency technologies in detail, including heat distribution systems, passive solar design (usually structural elements built into a building which access natural sunshine as heat) and passive cooling measures (again, most commonly integral design elements or add-ons that use natural airflows or protection from sunshine to minimise or eliminate electric-based cooling) which are relatively simple and low cost.

What Exactly is Energy Efficiency?

Energy efficiency can be described as using less energy to provide the same level of energy service. This might be achieved through technological changes like putting insulation in a house so that less energy is needed for heating and cooling to maintain a comfortable temperature. It can also be achieved by non-technical means such as better organisation and management: for example, turning off appliances when not in use. However, as the World Energy Council (WEC) explains, some behavioural changes that reduce energy use are not necessarily considered to be energy efficiency measures. For example, where consumers face constraints imposed by high energy-prices, they may decrease their consumption through a reduction in energy services (driving less or turning down the heating). However, these reductions don't necessarily improve the overall energy efficiency of the economy and are easily reversible. On the other hand, if one makes energy efficiency improvements in a home or office through better energy management and behavioural changes, these can be understood as energy efficiency measures. If domestic or company energy policy changed for the worse, then it is possible to see how a slide backwards into increased demand for energy used for a given service might occur. This just goes to show how important energy management policy and building occupant-behaviour are as tools for energy efficiency.

Assuming that our homes and offices are on a general trend, if very slow, of improved energy efficiency, ideally any initiative aiming to improve a building's energy efficiency will encompass so-called software changes (non-technological means such as human behaviour and energy management policy) as well as hardware changes (technologies such as energy efficient devices and equipment). Energy efficiency also means: eliminating massive amounts of wasted energy, staying warm in winter, saving significant sums of money and helping to create a more sustainable world. Using less can be associated with software changes such as taking less baths or shorter showers and remembering to turn the lights off when you leave a room. Technical efficiency, or hardware, changes would be fitting a new more water-efficient shower head and replacing incandescent bulbs with compact fluorescents that use 25 per cent of the energy for the same amount of light. In an office situation, the latter might be supplemented with occupancy sensors to turn lights on and off according to whether there's someone in the space or not (more on this below).

Energy Efficiency Across the World

Energy efficiency means something different in the various climate zones across the world. Even within the US, more efficient space cooling and air-conditioning can be the main objective in Miami while winter heating is the bigger issue in New York. The same is true in Europe, where southern Spain has to deal with extreme, close to African, heat while Finland suffers very cold, almost Siberian, winters. All over the world, significant energy saving potential remains untapped with techniques existing that could cut existing buildings' consumption by 50 per cent or more, but with the building renovation rate very slow. Furthermore, it is not difficult to reduce the energy use of many electrical appliances but, again, the uptake has been slow. This is almost certainly a consequence of a global economy that has undervalued its energy supplies.

The US and Energy Efficiency

In the US each state and each municipality has its own way of regulating energy efficiency measures, while the Federal level offers guidelines and sometimes incentives. The most stringent regulations are generally found at the local house owners' association level. The US Department of Energy also runs an Energy Efficiency and Renewable Energy (EERE) initiative. This has a mission to enhance energy efficiency and productivity and bring clean, reliable and affordable energy technologies to the US marketplace. The Energy Star Program, developed with the US Department of Energy backing, was introduced as a voluntary program in 1992. It provides technical information and tools that consumers can use to select energy efficiency solutions and best management practices, making it possible to save money while protecting the environment.

The UK and Energy Efficiency

The energy consumed in UK homes is responsible for about 25 per cent of the UK's CO_2 emissions, with much of the energy being lost to waste in aged housing stock designed for coal fires which need draughts to work, rather than twenty-first-century centrally-heated spaces. As part of the EU, the UK depends greatly on a network of Energy Advice Centres and Energy Agencies. Over 20 of the local energy advice centres are managed by the Energy Saving Trust (EST) which specialises in energy solutions at the household and community levels. For businesses there are the Carbon Trust and Energy Agencies. All these organisations have comprehensive websites plus free-phone numbers to call for advice. The Carbon Trust has a large team of energy advisers who can answer telephone enquiries or make site visits which can involve full energy audits. Both the EST and the Carbon Trust can also advise on the availability and range of grants and incentives.

Europe and Energy Efficiency

Energy efficiency is at the heart of the European Union's Europe 2020 Strategy, focused around an Energy Efficiency Plan which was ratified in March 2011. The EU recognises that energy efficiency is one of the most cost-effective ways to enhance the security of energy supply, and to reduce emissions of GHGs and other pollutants. The new 2020 strategy aims for smart, sustainable and inclusive growth and a transition towards a resource-efficient economy. In many ways, energy efficiency is Europe's biggest under-tapped energy resource. The EU has set a 2020 target for saving 20 per cent of its primary energy consumption compared to projections – an objective identified in the Commission's Communication on Energy 2020 as a key step towards achieving our long-term energy and climate goals. One advantage of using 20 per cent less energy will be diminishing the need for additional power-generating plants – a clear saving. However, despite substantial steps having been taken towards the 2020 objective – notably in the appliances and buildings markets, recent estimates suggest that the EU is on course to achieve only half of the 20 per cent objective. To deal with these issues, the EU has developed a comprehensive new Energy Efficiency Plan. With around a quarter of final European energy consumption being for domestic space heating, all Europeans have a role to play in reducing energy use.

Energy Efficiency Measures

Most of the measures in Figure 2.2 and applicable to the average home are also relevant to many business environments, be they offices, studios or workshops. The following section takes a closer look at measures more specifically appropriate for business environments.

Figures 2.1a and 2.1b
Average home energy use in the USA and UK. In the UK the largest use is space heating alone, while in the USA, with its higher temperatures in many places, heating and cooling are the biggest energy users.

Sources: (a) Dan Chiras (2006), 'The Homeowner's Guide to Renewable Energy', New Society, British Colombia; (b) Centre for Alternative Technology, www.cat.org.uk

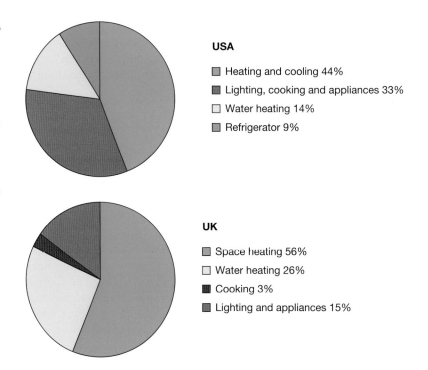

USA
- Heating and cooling 44%
- Lighting, cooking and appliances 33%
- Water heating 14%
- Refrigerator 9%

UK
- Space heating 56%
- Water heating 26%
- Cooking 3%
- Lighting and appliances 15%

Keeping Buildings Cool

Keeping our buildings comfortable to live and work in without using renewable energy sources brings with it an enormous cost in terms of household or business finance, fuel and GHG emissions. The same is true whether we are cooling a space in hot weather conditions or heating it up in cooler climates. Whether energy is being utilised for cooling or heating purposes, there are some basic energy efficiency measures that can be applied.

The cooling and air-conditioning of rooms and whole buildings can be a bigger energy issue than heating. The US Department of Energy (DOE) claims that the average US home spends more than $2,200 a year on energy. Heating and cooling costs the average homeowner around $1,000 a year – nearly half the home's total energy bill. If your central air-conditioning unit is more than 12 years old, they say that replacing it with an Energy Star approved model could cut your cooling costs by 30 per cent. The US DOE website also offers online advice on hiring contractors and selecting energy rated air-conditioning units. In the UK, the Carbon Trust also offers detailed online information.

Thermal comfort levels generally mean that we want to keep internal spaces at around 70°F to 75°F (21°C to 24°C) even on hot summer days. When it comes to keeping a building cool, the main factors to consider are air-flow, insulation and shading. In this chapter we focus mainly on passive-cooling techniques. Active methods, such as the use of heat pumps, are touched on below but covered in detail later in the book.

Airflow

Natural ventilation is always the cheapest and most simple method, particularly if it can be designed for a new build. If you live in a climate that has to strike a balance between hot summers and cold winters, draught-proofing or caulking of a property (see Keeping Buildings Warm, below, for more on this) can provide double benefits: greater efficiency in both cooling and heating. Simply by holding the cool or hot air in a building for longer makes cooling or heating cheaper on the pocket and the environment. Mechanical ventilation is also commonly used but is not a passive technique. The simplest form is a room fan but others, such as earth tubes, are increasingly common options. The earth tube concept is to bury tubes deep enough in the ground to access an even year-round temperature, and then to pass air from the outside through the subterranean tubes to

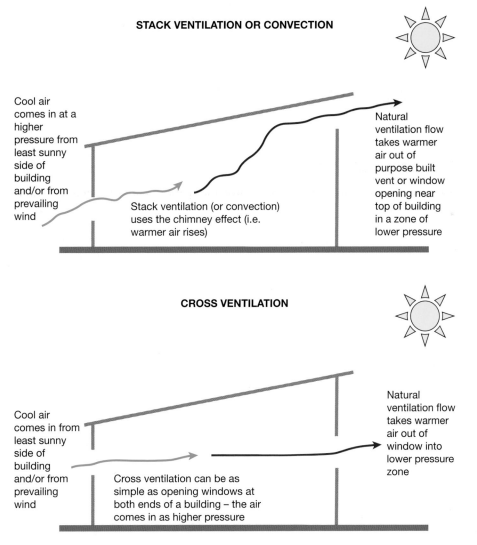

Figure 2.2a and 2.2b
The two main types of natural ventilation: (a) stack (or convective) ventilation and (b) cross ventilation. Natural ventilation can create greater comfort and savings, particularly where cooling is required.

Source: Dilwyn Jenkins

be heated or cooled before entering the building (in the summer the earth is cooler than the outside air temperature so air will be cooled as it goes through the tubes and vice versa in the winter).

Insulation

Installing insulation can add to the efficiency of a cooling system by keeping the cool air inside a building cool for longer, just like a cool box which is lined with insulating material to keep ice packs frozen for long periods. In particular, the insulation of thermal mass (generally in or on the walls or ceilings of a property) can significantly reduce active cooling needs. See the section below on Keeping Buildings Warm, for more information on insulation.

Shading

Cooling can be made easier and more efficient by shading areas of both the external walls and roof, as well as windows, to reduce the amount of heat coming into a building. Externally, a building can be insulated or shaded (which is often easier, e.g. by trees and other vegetation, or well-positioned screens located externally). Up to around 50 per cent of the heat in a room can be created by the sun through windows, so appropriate use of different materials can help reduce this simply by shading glass windows from direct sunlight. This kind of shading is often built into office buildings, although some companies have found the need to install their own shading or sun-blinds after the original design and construction, even in ultra-efficient, purpose-built premises. Wooden shutters, cloth roller blinds

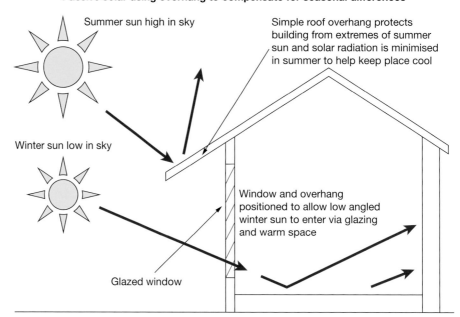

Figure 2.3 Simple structural devices like roof overhangs can be designed to minimise solar gain in summer and maximise it in winter

Source: Max Jenkins

and high-tech netting (like mosquito netting or Phifer Suntex) can be very effective. Shutters and netting are usually fixed externally, while blinds tend to be an internal feature. Trailing or climbing plants can also be used to minimise solar gain through windows.

Other Techniques

Reflective roofs and walls, night ventilation, glass that lets light in but keeps heat out, use of phase-change materials, reflective screens on the inside of windows (e.g. Reflectix material), water features and indoor plants are all commonly utilised methods of reducing heat gain for direct sunlight.

Many models of heat pumps (see Chapter 5) can be reversed and used for cooling, which can be useful in regions with extremes of climate and temperatures. They function by absorbing heat from the outdoor air in winter and expelling heat into outdoor air in summer. Liquid refrigerant passes through the expansion device, changing to a low-pressure liquid/vapour mixture before going to the indoor coil, which acts as the evaporator. The liquid refrigerant absorbs heat from inside the building before expelling it as a hot gas via the outside coil which works as a condenser.

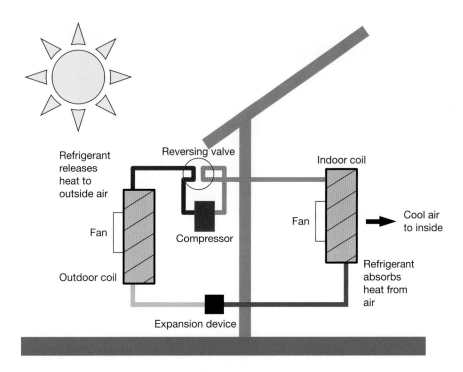

Figure 2.4 Workings of an air source heat pump operating in cooling mode

Source: Dilwyn Jenkins

Keeping Buildings Warm

Figure 2.5 Optimising natural light inside building and helping to heat the space by maximising solar gain via extensive surface areas of glass (in walls, windows and skylights) at the Centre for Alternative Technology, Wales, UK. The wall to the left provides thermal mass heat storage, being made of rammed earth.

Source: Dilwyn Jenkins

Air-Related Measures

Draught-proofing is one of the simplest energy efficiency measures to introduce to home or office spaces, so there's no excuse for anyone suffering discomfort from indoor breezes. Nevertheless, it is important to ensure that there is always sufficient ventilation for the building or room's heating system to operate safely and efficiently. The correct balance between these two factors is the key to successful draught-proofing.

Draught-proofing materials vary slightly between countries to suit local styles of windows and doors. Essentially they are all designed to offer a draught (and driving rain) seal. It's a matter of choosing, from a local or online store, the material that best suits your particular door and window frames. As a general rule of thumb, they should be fixed to internal or external frame surfaces with the door or window closed. More often than not, doors will have seals fitted externally and windows to internal surfaces. In extreme cases you might want to proof internally as well as from the outside.

A common part of the solution to household draughts has long been the use of curtains to cover the whole window area or more (ideally with an insulating backing to the curtain). This is cheap and effective and can be used for doors and animal flaps, too. Similarly, using stuffed cloth sausage dogs (which can be bought or easily made from recycled cloth and filled with beans) at the base of external or internal doors can significantly reduce draughts. A thorough testing of airflows and airtightness (air changes per hour or ACH) in a building can be achieved by a blower door test.

Some tests also involve an infra-red or even a PFT (PerFluorocarbon tracer gas which is excellent for longer-term analysis) air infiltration measurement to get to grips with air leakage/infiltration. A fairly airtight modern house is likely to achieve as little as 0.6 ACH. A simple alternative method of finding unwanted airflows involves smoking out the draughts (e.g. with an incense stick on a windy or cool day) room by room, window by window, crack by crack. Generally speaking, the newer the house or business premises, the less air changes per hour.

Key places to search for and eliminate draughts are listed below.

Door frames

Significant draughts enter buildings through external doors. These should be proofed with the appropriate range of seals and strips. Compression seals and

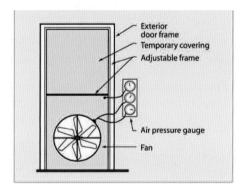

Figure 2.6a Testing the air tightness of a home using a special fan called a blower door can help to ensure that air sealing work is effective. Often, energy efficiency incentive programs such as the DOE/EPA Energy Star program require a blower door test, usually performed in less than an hour, to confirm the tightness of the house.

Source: The Office of Energy Efficiency and Renewable Energy (EERE), www.eere.energy.gov

2.6b A blower door in situ. Door blower tests use a powerful fan to pull air out of a building to create exaggerated but detectable drafts.

Source: Retrotec Inc., www.retrotec.com

brush strips work well on all but the most irregular of doorways. Wiper seals made with nylon brush or rubber blades work well on door sides and most windows.

Window frames

External doors will always benefit from draught proofing and sometimes it's good to proof internal doors, depending on the use of spaces in the building. Ensure catches close properly and fit wiper or small compression seals to the appropriate internal closing surfaces. Fill any gaps.

Chimneys

If a chimney is in use, make sure the chimney or flue has a damper mechanism to close it off when not being used. For chimneys no longer in use, it's best to board them up (e.g. with an insulating or heat-proofed board). It is also possible to stuff a chimney with fabric, sackcloth or foam offcuts. Some people inflate a sturdy balloon in the chimney, making a compression seal.

Loft hatches

A loft hatch is easily made airtight with compression seals along all edges and a catch that brings the hatch door into compression. It can be topped up with 50mm of insulation, loft side.

Letterboxes, keyholes and animal flaps

Brushes or metal/plastic flaps can be fitted to the inside, or something similar can be fabricated from an old scrap of carpet or rubber, cut to the right size and shape and pinned along the top edge.

Figure 2.7a and 2.7b External and internal photograph of well draught-sealed letterbox with brush in the middle and sprung flaps on both sides

Source: Dilwyn Jenkins

Gaps in floors, around pipes, cables, electrical switches, recessed ceiling lights, skirting boards and ceilings etc.

These can be filled with the appropriate sealant (silicon, papier maché, cork, powder-based filler or a fibre filler, like oakum, made from hemp or jute) for small gaps in the floor and skirting boards, windows and door frames etc. Expanding foam can be used for larger gaps (the best is non-porous closed-cell foam blown with pentane rather than gases that are ozone depleting or GHGs), applying an internal and/or external finish when foam has hardened off. Loose-fill insulation can also be used (e.g. to fill gaps behind pipes).

Internal to external ducts

Fans extracting hot, smelly or damp air or fumes should have or be fitted with self-closing covers for when they are not in use.

Insulation

Insulation is fundamental to energy efficiency, particularly when it comes to keeping spaces warm or cool or heating water. We all know that heat transfers and moves from warm spaces into cooler ones. On a cool day around 40 per cent of a home's heat is lost just through the average floor and loft, with walls losing a further 35 per cent. When it's hot outside the opposite is true – heat passes through the windows, walls and ceilings and can make interiors uncomfortably warm. Thermal insulation slows down this transference of heat and can help keep buildings warm or cool, as required. Insulation has a good financial rate of return

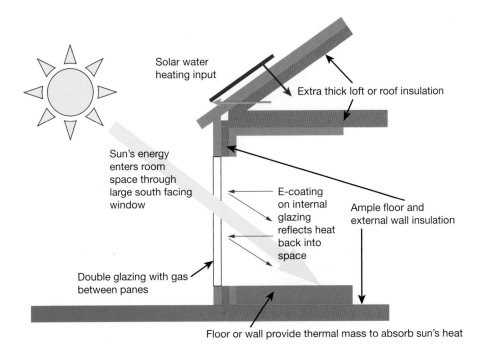

Figure 2.8 This diagram illustrates a number of simple and standard techniques for keeping an indoor space warm: insulation (in ceiling, wall, floor and loft), e-coated double glazing and natural solar gain through windows.

Source: Dilwyn Jenkins

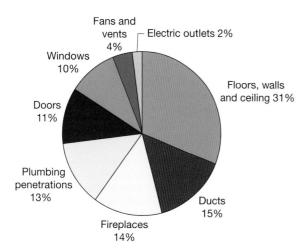
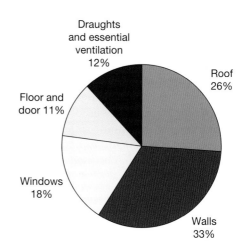

Figure 2.9a (left) In a cool climate, the majority of heat lost from an uninsulated, free-standing house is through the walls, followed by the roof. However, the roof has a much smaller surface area than the walls indicating that more is lost per square foot from the roof. In terms of installing insulation, this is usually where the biggest and easiest savings can be made.

Source: Energy Saving Trust, www.est.org.uk

Figure 2.9b (right) Air infiltrates into and out of your home through every hole and crack. About one-third of the air infiltrates through openings in your ceilings, walls and doors

Source: Department of Energy, www.energysavers.gov

with hot-water tank jackets famously having the fastest payback: they cost next to nothing and, if you're using mains electricity as the heat source, they can cover their cost twice over on the first electricity bill. According to the Energy Saving Trust, if everyone in the UK fitted a hot-water cylinder jacket that was at least 75mm thick, there would be enough carbon dioxide saved per year to fill seven million double decker buses!

DIY energy audits can usually pick up many problems in any property. By keeping a checklist and inspecting room by room as well as outside, it is possible to get an accurate picture of where energy is used, and lost, and also to prioritise any energy efficiency improvements that may be needed. In terms of making the most impact for the smallest investment, the first considerations should be drafts and air leakage (remembering that some ventilation is required for a healthy house), then insulation measures to minimise heat loss through ceiling, walls and floor. In the US, the DOE gives minimum recommended insulation values in terms of R-Values according to six nationwide 'insulation zones'. Most US houses should have between R-19 and R-49 insulation in the attic (6–15 inches mineral wool insulation). In the UK, the recommended depth for mineral wool insulation is 270 mm, though other materials require different depths. The insulation levels of exterior walls are best judged by their material constituents and construction.

Passive Solar Design

In a home or office, incorporating passive solar design elements can make a big difference to a building's energy consumption. Passive solar buildings range from being heated almost entirely by the sun to properties that have (in the Northern Hemisphere) south-facing windows able to provide a portion of the heating load. The trick is taking the best possible advantage of your particular location and climate. Natural ventilation, for example by opening windows in a space that's too warm and not being heated or air-conditioned, can form part of a simple solar passive design. Again, this is simply taking advantage of free natural processes that may be specific to your building and locality.

Size and orientation of windows

In the North, the south-facing side of a building should have more window surface area (the opposite for cooling and the reverse in winter for countries in the South such as Australia and New Zealand). These windows should be kept clear and clean.

Shading controls (interior and exterior)

This includes the strategic placing and use of blinds, drapes, shutters, roller screens etc. Keep the (south facing) blinds closed during the cooling season to reduce solar gain, but open in the day during heating season to maximise solar gain.

Sun pipes

These are tubes that bring natural light into parts of the house where it is needed and are generally more energy efficient than adding windows or skylights.

Phase changing materials (PCM)

These take advantage of phase changes associated with temperature differences in certain materials (e.g. some plasterboards are available with a wax core: the wax melts and naturally absorbs heat as the board temperature rises, then yields heat back to the room as it cools).

Solar absorber surfaces and building mass

Materials such as concrete, stones, brick, tiles and even water have been successfully used for thermal storage and slow release of heat (generally after sunset).

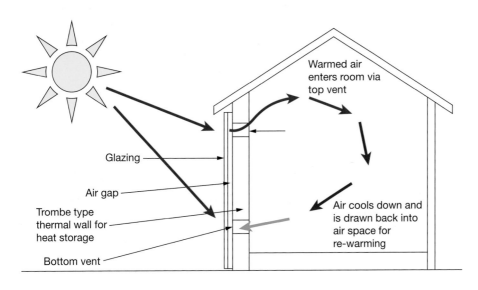

Figure 2.10 Extended surface areas of glazing and Trombe type walls can be used for optimising solar gain and an element of heat storage

Source: Max Jenkins

Figure 2.11 An attached conservatory or sun room can be used effectively to reduce heat loss and provide a naturally sun heated space

Source: Max Jenkins

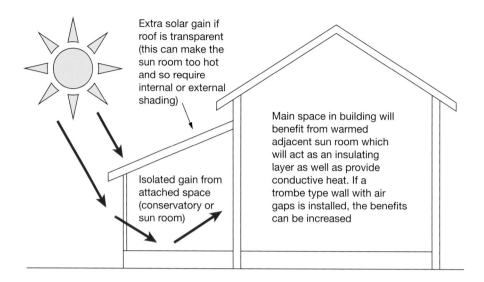

Passive House Planning Package (PHPP)

Developed in Germany by the PassivHaus Institut, PHPP is a software package for architects and designers to model the thermal, energy and carbon performance of a property with a view to selecting the best available and most energy efficient technologies, including the airtightness, insulation, heating system, appliances and passive design elements. The software can be bought online. The package works very well for new build and can be applied effectively to refurbishments.

> ### Thermal conductivity, R-values and U-values
>
> Thermal conductivity is a measure of how fast a material conducts heat. In relation to buildings, all building materials, from concrete to glass, have been tested to measure how fast heat will pass through them over a period of time. Thermal conductivity (K-value) is for a unit area of a material of unit thickness for a unit temperature. The reciprocal of thermal conductivity is thermal resistively (R-value): a measure of how resistant a material is to heat passing through it. Two other values are also used in relation to heat transfer and these refer to materials of a particular thickness: R-value for thermal resistance and U-value for thermal transmittance. U- and R-values are often used in comparing one building product with another. For example, a single-glazed window will have a lower R-value (and therefore a higher U-value) than a double-glazed window, showing that it transfers more heat and so is less suitable for keeping the heat where you want it. Even more efficient would be an argon-filled, low-e double-glazed window where the gap between the specially coated glazing is filled with argon gas. This would have much higher R-values (lower U-values) and would be the most effective insulator of the three.

> In the US the measurement units for distance are based on feet, BTU and Fahrenheit and R-values tend to be figures that are used. The Federal Trade Commission (FTC) is the body that governs claims about R-values in order to protect consumers.
>
> In the UK and most of the rest of the world measurements are based on the International System of Units (SI) and use metres, Watts and Kelvin or Celcius. U-values are the figures more commonly used in the UK.
>
> For conversion purposes between US and SI figures, the formula 1 W/mK = 1.730735 Btu is used. US and SI R-values can be confused when cited without specifying the units (e.g. R-3.5).
>
> Thermal conductivity (represented as either 'k' or 'psi') is a measure of how fast a material conducts heat.
>
> R-value is quite different to thermal conductivity (and reciprocal of the U-value: see below) since it measures thermal resistance in buildings and building materials. They are usually stated as the ratio of temperature difference across an insulator and heat-flow per unit area. Higher R-value numbers, and lower U-values, both represent more effective insulation. U-values represent the overall heat-transfer coefficient, which means that they describe the rate of heat transfer through a building material over a given area. Recommended U-values (or factors) describe a recommended maximum, while recommended R-values would specify a minimum.
>
> U is the inverse of R with units of Watts, metres and Kelvin (or Celsius) rather than the:
>
> - R-value = thermal resistivity (ft$^2 \times$°F/Btu, m$^2 \times$K/W or m$^2 \times$°C/W)
> - U- value = thermal conductivity properties (1 h$\times$ft$^2 \times$°F/Btu or W/m^{2}K)
> - R = thermal resistivity (ft$^2 \times$°F$\times$h/Btu, m$^2 \times$K/W or m$^2 \times$°C/W)
>
> Heat loss per square metre is calculated by dividing the temperature difference between inside and outside of the materials by the R-value. Doubling the thickness of fibreglass insulation effectively doubles its R-value (e.g. from 2 m^{2}K/W for 110 mm thick material or 4 m^{2}K/W for 220 mm of the same material). Additional encapsulating layers over the insulation material marginally adds to the R-value.
>
> R or U-values of insulating materials are usually rated by the material's manufacturer under test conditions.
>
> With loft insulation, if the temperature inside the house is 22°C and 11°C in the roof space, the temperature difference is 10 C° (= 10 K). If the ceiling insulation is rated at R–2 (R = 2.0 m^{2}K/W), energy will be lost at a rate of 10 K / 2 K·m^2/W (i.e. 5 watts for every square metre of ceiling).

Insulation Materials

For the sake of sustainability, and throughout this book, we refer mainly to insulation materials that lock carbon up in the fabric of the building rather than those that generate damaging GHG emissions during their lifecycle (e.g. during primary sourcing, manufacture and delivery). Fibreglass and rock wool, for

Figure 2.12a A pack of sheep's wool insulation – a natural fibre insulation for the healthy home

Source: Dilwyn Jenkins

Figure 2.12b A pack of hemp insulation

Source: Dilwyn Jenkins

Figure 2.12c A sample block of rigid foam insulation with reflective backing for extra thermal benefits

Source: Dilwyn Jenkins

Figure 2.12d A sample block of thick cork insulation

Source: Dilwyn Jenkins

instance, have significantly lower embodied energy (and associated emissions) than standard foam insulation materials. Natural fibre insulation materials – cellulose, fibreboard and hemecrete for instance – have even less impact.

Things to consider when buying insulation materials include:

- thermal performance (R-values or U-values);
- financial cost;
- environmental cost (embedded energy and potential as pollutant or irritant);
- expected effective life;
- ease of installation (choosing materials that will most easily resolve the problems to be solved in a specific home or business premises);
- resistance to fire, pests, moisture or compression; and
- ease of recycling or disposal at end of useful life.

The US DOE recommends first checking the insulation in your attic, ceilings, exterior and basement walls, floors and crawl spaces to see if it meets the levels recommended for your area. They recommend ranges of R-values based on local heating and cooling costs and the climate conditions in different areas of the nation and their website has an online Zip-code Insulation Calculator, based on weather data for where you live and basic information on the property in question. State and local code minimum insulation requirements may be less than the DOE recommendations, which are based on cost effectiveness. They classify insulation materials as coming in four main types, each with different characteristics: cotton, sheep's wool, straw bales and straw boards/panels.

Table 2.1 Natural fibre insulation materials

Material	Details	R-values (US) per-inch
Cotton	Consists of 85% recycled cotton and 15% plastic fibres that have been treated with borate – the same flame retardant and insect/rodent repellent as cellulose insulation. One product uses recycled blue jean manufacturing trim waste. As a result of its recycled content, this product uses minimal energy to manufacture. Cotton insulation is available in batts with a thermal resistance, or R-value of R-3.4 per inch. Cotton insulation is also non-toxic; you can install it without using respiratory or skin exposure protection. However, cotton insulation costs about 15%–20% more than fibreglass batt insulation.	3.4
Sheep's wool	For use as insulation, sheep's wool is also treated with borate to resist pests, fire, and mould. It can hold large quantities of water, which is an advantage for use in some walls, but repeated wetting and drying can leach out the borate. The thermal resistance or R-value of sheep's wool batts is about R-3.5 per inch, which is similar to other fibrous insulation types.	3.5
Straw bales	Straw bale construction, popular 150 years ago on the Great Plains of the United States, has received renewed interest. Straw bales tested by Oak Ridge National Laboratory yielded R-values of R-2.4 to R-3.0 per inch. But at least one straw bale expert claims R-2.4 per inch is more representative of typical straw bale construction due to the many gaps between the stacked bales.	2.4
Straw boards/ panels	The process of fusing straw into boards without adhesives was developed in the 1930s. Panels are usually 2–4 inches (51–102 mm) thick and faced with heavyweight kraft paper on each side. Although manufacturer's claims vary, R-values realistically range from about R-1.4 to R-2 per inch. The boards also make effective sound-absorbing panels for interior partitions. Some manufacturers have developed structural insulated panels from multiple-layered, compressed-straw panels.	1.4–2
Hemp	Hemp insulation is relatively unknown and not commonly used in the United States. It offers a similar R-value as other fibrous insulation types.	3.5

Source: US Department of Energy, http://energysavers.gov

Figure 2.13 Sheep's wool insulation bats with the source of their raw material

Source: Thermafleece, www.thermafleece.com

Figure 2.14a and 2.14b Sheep's wool insulation being fitted into timber-framed building wall-panels at the Centre for Alternative Technology, Wales, UK

Source: (a) Pat Borer (b) Centre for Alternative Technology, www.cat.org.uk

Figure 2.15 Photograph showing tan-coloured internal wall which has been insulated and re-plastered simultaneously with a lime and hemp mix

Source: Pat Borer

Figure 2.16 Photograph of dry-lined wall filled with flax insulation rolls

Source: Pat Borer

Rolls and batts

These are flexible products like sheep's wool, fibreglass and rock wool and usually available in widths suited to the standard spacings of wall studs and attic or floor joists.

Loose-fill insulation

This is usually rock wool, or cellulose in the form of loose fibres or fibre pellets (often blown into spaces where it conforms readily to cavities and structural forms; ideal for places where it is difficult to install).

Rigid foam insulation

This is typically more expensive than fibre insulation and much more costly in environmental terms. It is nevertheless useful in buildings where high R-values are required and where there is little space available (R-values range from R-4 to R-6.5 per inch).

Table 2.2 A rough guide to the type and effectiveness of various insulation materials

Insulation material	Form	R-value per inch	Typical applications
Cellulose	Loose-fill	3.2	Walls & lofts
Cotton	Rolls & batts	3.4	Walls & lofts
Fibre glass (an average)	Rolls & batts	2.6	Walls & lofts
Hemp	Rolls & batts	3.5	Walls & lofts
Rock wool	Rolls & batts	3.1	Walls & lofts
Sheep's wool	Rolls & batts	3.5	Walls & lofts
Straw	Rolls & batts	1.4 to 2	Walls & lofts
Expanded polystyrene	Rigid board	3.8 to 4.4	Floors, walls & lofts
Icynene	Expanding foam	3.6	Walls & lofts

Source: Dan Chiras (2006) and US Department of Energy (DOE)

Similar information is available in the UK from the EST website (see Annex 2), which also offers an online search tool to help householders find registered local suppliers and fitters of insulation materials as well as detail on cavity wall, solid wall, floor and loft insulation, draught-proofing and glazing.

Figure 2.17 External insulation (white) being applied to existing wall, Berlin, Germany

Source: Frank Jackson

Insulation and moisture

It's important to keep most types of insulation materials from absorbing moisture, particularly in wall cavities where it might be difficult to notice it accumulating over time and which can cause health problems to both building fabric and its occupants. Cavity insulation is most commonly protected with a vapour barrier: a plastic-sheet layer often applied to cavity stud walls during construction. If these are damaged or not there at all, they should be fixed or installed after filling the cavity and before finishing internal surfaces.

Cavity wall insulation

It only takes a few days to install, but it is a job most householders and businesses would hire a professional to execute. The usual technique involves drilling holes into the cavity from the outside wall or inside drywall. In a timber-framed house the holes aim to access each stud cavity. In a brick-built house the cavity spaces are generally larger. Internally, if you have stud walling it may be possible to increase the depth of the cavity (and consequently the insulation material) by removing the plasterboard or drywall and adding another layer of frame (be careful how this impacts on window and door frames).

Once the appropriate holes have been drilled the insulation material can be blown in (usually of a cellulose variety). The advantages of getting a professional to do the work range from the availability of the right equipment and experience with different building types to knowing how to make sure the cellulose doesn't form air pockets by getting trapped on pipes and wires hidden in the cavity. In North America, licensed companies can fill wall cavities with a water-soluble and relatively environment-friendly foam insulation material – Icynene – which contains no formaldehyde or ozone depleting chemicals.

Solid wall insulation

The usual practice is to insulate internally either under the drywall, between the studs, or by increasing the depth of cavity by adding new studs on top of the existing ones (as above with cavity walls), then fixing (or blowing) insulation into place before sealing up with a new false wall.

External wall insulation

Covering the outside surfaces of a home or office building can achieve excellent U-values, particularly if the premises is complemented with all the other possible forms of insulation. The insulation material is covered with a waterproof top decorative layer (or render) which can be painted. The most difficult part of this work is ensuring that sufficient insulation finds its way to the door and window reveals as well as behind pipes, under gutters, overhangs etc.

The least expensive option is wet external rendering with insulation materials, but dry cladding – basically adding a supporting framework to fix cladding to – has the advantage of permitting access for routine inspection. Depending on the type of insulation material, the dry clad system will probably require a vapour barrier.

For lofts and roofs

This is generally one of the easier forms of insulation to fit because many homes and office buildings have loft spaces or attics where batts of material can simply be unrolled. The recommended depth of rock wool insulation in the UK is presently 270mm (close to one foot) but it is always sensible to go one step further than national recommendations, not least since they are themselves being constantly upgraded. If this isn't possible – for instance if the attic space is used as an office or bedroom and the ceilings are sloped with the shape of the roof – then the insulation should be fixed between the rafters before fixing a vapour barrier and final board and/or plaster layer.

Roofs can be insulated from the outside only at the same time as replacing the roofing cover (slates, shingles etc.). At this time, however, it's simple enough to fit batts of insulation between vapour membranes, before re-covering the roof. Natural ventilation is important for loft spaces. As a minimum, roof spaces with a roof pitch of less than 15° need a continuous 25 mm gap around the eaves for ventilation (a pitch of over 15° needs a 10 mm gap). Fixed vents should be covered with insect-proof mesh.

For ground floors

Concrete floors can benefit substantially from insulation below the concrete cap (i.e. during construction or floor re-laying). If there's not enough space you can raise the floor levels.

If there is space, insulation placed between the concrete slab and external wall surfaces can stop cold bridging. Fitting underfloor heating pipes within the top concrete layer or beneath it, on top of the insulation, can also be a very effective way to heat home or business spaces.

Sub-subterranean basements might benefit from insulation under the bottom concrete floor and also fixed between the ceiling joists between the basement and ground floor. As a minimum, basements should be sealed off with insulation between ceiling joists and attention paid to any potential cold bridging.

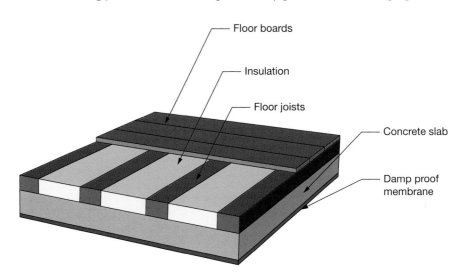

Figure 2.18 If it's not possible or appropriate to put insulation under the concrete floor, then an insulation layer can be laid on top, either between floor joists for a wooden top-surface, as shown here, or under a second layer of concrete

Source: Jacinta MacDermot

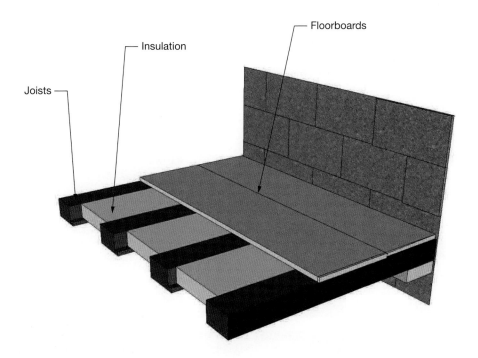

Figure 2.19 When it comes to suspended wooden floors, the boards need to be lifted and the insulation laid between the joists and, ideally, filled to the full depth of the available space. Vapour barriers should not be fitted in case of liquid spillage.

Source: Jacinta MacDermot

Windows and doors

Upgrading windows and doors can typically achieve reductions in heat loss of up to about 4 to 8 per cent. This is less than can be achieved with draught proofing or wall and loft insulation. Large single frame and metal-framed windows can be particularly inefficient in retaining heat within a building.

Figure 2.20 Double-glazed window unit with gas-filled space between the two glass panes

Source: Dilwyn Jenkins

If the windows and doors are not due for replacement – or are bound by special planning considerations (e.g. historic building or town centre) – there are other more simple options such as secondary glazing and storm shutters. However, there are plenty of high-performance windows available on the market and most possess built-in draught and weather proofing.

Storm windows (external)

Figures 2.21a, 2.21b, 2.21c, 2.21d External storm windows can protect the window sashes and original glass, as well as reducing energy loss. Often they can be opened, although not in the case of the window in 2.21b, fitted to Chapel House in Great Limber, UK.

Sources: Storm Windows Ltd, www.stormwindows.co.uk and Larsons.

Double or triple glazing (internal)

These help keep warmth in during winter as well as heat out during summer. Double-glazing with gas (usually argon, krypton or krypton) injected into the vacuum between the panes helps improve R- and U-values. If there's no injected gas, the panes should be separated by a gap of over 15 mm. Insulated spacers – between the window frame and the edges of the glazing – helps reduce heat loss and also moisture condensation. Some windows these days have the internal glazing surface e-coated (an invisible layer of tin or silver oxide which slows down heat loss through the glass), which reflects some heat back into the room rather than allowing it to leave via the window.

Secondary glazing with glass panes or clear plastic sheets is obviously much less expensive than replacing whole windows. This is usually fitted as a semi-permanent feature with a wooden, light metal or plastic frame, designed to be easily removed at the end of winter and repositioned at the start. Cheaper than standard glass or rigid plastic secondary glazing, but less durable and generally less effective, the basic form of secondary glazing is a thin plastic (similar to cling film) which can be fitted around a window or specific panes with magnetic clasps or double-sided sticky tape. With this latter technique, once fitted the wrinkles can be smoothed out with hot air from a hair dryer.

Manufacturers' information contains details of R- or U-values and air-infiltration rates. In the US the energy performance of fenestration products (doors, windows, skylights etc.) is rated and certified by the National Fenestration Rating Council. In the UK, the British Fenestration Rating Council (BFRC) uses rating levels with the highest being 'A' and with 'C' being the minimum recommended for an eco-home. Solar gain – the heat energy from the sun that is gained inside a building (expressed as a 'G' value on BFRC associated marketing labels) – is another issue for windows. Ideally this is maximised in winter and minimised in summer, otherwise it can cause overheating. Solar gain depends to a great measure on building location and orientation. If it's likely to be a major issue this should be discussed with your window supplier or installer, flagging up the particular conditions for your property, or specific windows in a given building. The alternative is to consider options for shading and even shutters at certain times of year.

Doors

As well as fitting better, a new door will ideally have a layer of insulation material sandwiched between the outer and inner surfaces. The door-frame should be insulated against cold bridging and fully draught sealed when closed. If the door has glazing, it should be high performance. There's a wide range of door types available, it's just a matter of choosing the right style and comparing U-values.

Lighting and Appliances

Our indoor lighting and the wide range of appliances found in the typical modern home or office consume large amounts of energy over a year, so it is vital to find ways to minimise their use wherever possible. Gizmos like power monitors and voltage optimisers are available and can help significantly, but the real solution comes down to having more efficient appliances and using them more carefully.

Energy monitoring and voltage optimisation for home and business

Monitoring electricity within a building has been common to large organisations and companies for some time but is becoming more common these days even at household level. At the very least energy monitors make end-users more conscious of which appliances and activities or processes are the chief guzzlers of electricity. There are many types and brands available, some of them very cheap. In the US products called Smart Power Strips are designed to help you reduce the amount of electricity you use by conveniently switching off those appliances normally kept on standby. In the UK, the Owl remote control socket pack fulfils a similar function.

Voltage optimisation is a relatively new technology that is increasingly utilised by businesses to generate power savings. They work by adjusting the incoming voltage to a constant voltage, usually either 220 V or 120 V, and also protect the building's electrical circuits from power surges. A UK-based savings calculator is available online to find out how much your home or business might benefit from investing in a voltage regulator, and another operates a free evaluation

Figure 2.22 Using a Watt meter in a home or office can help identify which appliances are consuming lots of electricity and at what times, a good starting point for deciding how to invest in more efficient equipment or measures

Source: Frank Jackson

service to see how useful their voltage-regulator technology might be for your building, if you call or email. In North America the Quebec-based company Power Quebec specialise in voltage optimisation products.

Lighting

Lighting accounts for around 10 per cent of electricity use in the USA and between 7 and 27 per cent across Europe. Globally, 19 per cent of electiricty used is dedicated to lighting or average, so it's clearly important for householders, office workers or workshop staff to reduce any waste from lighting. There are two first steps in achieving this: change the light source to lower energy type and ensure lights are off when a room is not being used. Beyond this, the actual siting of lighting fixtures can reduce lighting electricity demand (e.g. desk, kitchen or workshop task lighting, providing light only where it's needed rather than illuminating a whole room when it's not needed all of the time).

Compact fluorescent light bulbs (CFLs) should first be fitted in places where the light is often left on (e.g. hallways, lobbies, kitchen). Over the years, light quality from CFLs has much improved and is now available in a range of tones, from daylight to warm. Well-used CFLs not only pay back their cost in savings during the first year of operation but they also have a longer life expectancy. LED lighting technology has come on significantly in the last ten years and can now provide viable low-energy lighting solutions.

Whatever the type of bulb, if every light has its own switch lighting levels are much easier to control. Even if a room or corner of the office is empty for only 15 minutes, energy can be saved by turning the light off and on at exit and entry.

Figure 2.23a Compact fluorescent light bulb (CFLs). These typically use around 75% less energy than a conventional incandescent bulb to provide the same level of lighting (in many countries incandescent bulbs are no longer allowed to be sold).
Source: Energy Saving Trust, www.est.org.uk

Figure 2.23b Light emitting diodes (better known as LEDs) are even more efficient than CFLs. They also last longer and do not give off heat.

For offices in particular, there are substantial savings to be made by installing occupancy and light-level sensors to operate light switches. Lights with a time control switch are also sometimes used in areas that will be occupied for only a short period of time. Outdoor lights can also be controlled with movement sensors (motion or photocell type) so that they are only used when needed.

Lighting for the office and workshop

Since we spend most of our daylight hours in the workplace, maximising the use of daylight in the office or workshop makes sense and saves cents. Both the scale and occupancy behaviour tends to be different in office or workshop spaces compared to the conventional home, so there are some additional considerations for lighting in the workplace. All businesses could, for instance, benefit from occupancy sensors for lighting. These will turn lights on when someone enters a room, including bathrooms, and off again when they leave, circumventing the need to rely on staff awareness to do the job. These have a payback of about two years. Many workshops or factory spaces use overhead electric lighting even when there's sufficient daylight and task lighting. To minimise this, many businesses have found daylight sensors to be useful, which will switch the lights off when there is sufficient daylight entering the workspace. These can pay for themselves in less than one year.

Office equipment

There's no substitute for having an office energy-manager, but nonetheless it is every worker's responsibility to try and follow a few basic guidelines for saving energy:

- Equipment using adaptors, such as laptops, etc., should be turned off at the plug when not in use (their adaptors will otherwise continue to draw some power).
- Only switch equipment (computers, printers, photocopiers, etc.) on when you actually need to use it.
- Ensure your computer goes quickly into sleep mode when not in use and turn off screen savers.
- Share one printer with a network of colleagues.
- Ensure all equipment and lights are switched off at night.

Refrigeration

A sizeable percentage of most home energy bills relates to refrigerator and freezer electricity consumption. We tend to forget about them, but they are always switched on. The first thing to check is the setting: most refrigerators should be between 3°C and 6°C and freezers between -18°C and -15°C. The cooler the room the more efficiently fridges and freezers will work. This means they cost less to run in a cool pantry than a warm kitchen and, where possible, should be kept out of direct sunlight. The heat transfer grill – usually found at the back of refrigerators – needs to be kept clean and have sufficient air movement around it to work at top efficiencies. Open fridges and freezers as little and for as short a time as possible and try to cover food inside since any additional moisture content makes a fridge work harder to maintain a set temperature. Never put warm or hot food straight into a fridge – let it cool down first. Fridges and freezers should be defrosted regularly and any defective door seals should be replaced. If you plan to buy a new fridge or freezer go for one with high energy-label ratings and no larger than required (empty fridges work harder than full ones). An A++ fridge in the UK uses 50 per cent less electricity than an A rated one of the same size, making payback relatively fast.

Washing machines, tumble driers and dishwashers

Heating water for washing generates the major part of a washing machine's energy demand. Selecting lower temperatures for washes and ensuring every load is full makes a big difference to overall energy consumption. These days most washing machines provide useful energy label ratings. Tumble driers are heavy energy consumers so their use, if at all, should be minimised if you have access to an outdoor clothes-line. Again, energy-label ratings are a good guide. Dishwashers are also high energy-consumers and washing up without them is usually a lot more energy efficient. Energy use can be reduced by only using dishwashers for full-load washes and allowing the dishes to air dry. These, too, are energy-label rated.

Cookers

Conventional electric cookers use a lot of energy and, where possible, should be avoided. Induction cookers have recently come onto the market and, although they run off electricity, they are faster and more energy-efficient than conventional electric cookers. They are also more expensive to buy. Induction hobs work well with saucepans that have a base with high ferrous-metal content: if a magnet sticks to the bottom of the pan, it will be suitable for an induction hob. Microwaves use less energy to cook, but they are not as versatile or efficient as a gas stove. There is a tradition in many countries of cooking on a wood stove, which often simultaneously heats household hot water and sometimes provides hot water for a wet heat-distribution system. Hydropower can be a useful renewable source of electricity for cooking (see Chapter 8). There are also biogas and solar cookers but these are beyond the scope of this book, being presently designed for external use in rural situations. It's good practice not to use oversized pans for cooking, and to keep lids on pans while in use and to cook on simmering heats rather than boiling whenever possible. Pressure cookers can save energy, and all new conventional cookers tend to be energy-label rated.

Stand-by loads

Also called ghost or phantom loads, the proliferation of appliances – videos, DVD players, music centres, TVs, computer screens, laptops, PCs, printers, alarm clocks, microwave cookers, mobile phones, video game centres etc. – has held back overall levels of energy efficiency in homes and offices. As appliances themselves became more efficient, so the quantity we use and the number with integrated stand-by functions has increased dramatically. Some of these pieces of equipment were so poorly thought through at the design stage that they lose basic programming (e.g. time and date on a video player) if the stand-by function is turned off by switching off the power supply at the plug. The best remedy is simply to ensure all stand-by equipment is turned off when not needed (e.g. when going to bed).

Ideally, a house could be fitted with a master or central switch to turn off and on all the household or office appliances (except for fridges, freezers and electric alarm clocks) simultaneously when leaving or entering a building or sector of a building. Properly sized extension cables or power strips with built-in protection (such as a polarized plug and receptacle, grounded terminals, fuse link, or perhaps a residual-current device) can be used to perform the same function, but are not recommended for reasons of safety even for indoor applications (see section on Energy Monitoring above).

Ventilation, cooling and heating

Typically, ventilation, air-conditioning and heating systems control a property's humidity, air quality and temperature to maintain specific selected and programmed conditions. They work by moving heat and moisture between the inside and outside. Each type of system is available in a wide variety of scales and configurations, some for whole building control and others just for specific rooms. These systems are often assessed together because their functions are interrelated and they need to be programmed to work harmoniously (e.g. it would be very inefficient to be cooling and heating the same space at the same time).

Arguably, the simplest and most important aspect for a building user to get right is the setting of the electronic controls, and automatic (optimum) kick-in and kick-out options, for the temperatures and timings of ventilating, cooling and heating equipment.

In exceptionally well-designed, well-insulated and draught-proofed buildings the heat from appliances, people and solar gain can provide a significant proportion of the heat needed to keep occupants warm. In these circumstances, it may be possible to top up the heating successfully with some kind of heat recovery system, taking warmth from the outgoing air and recycling it as heat. For the typical home and office this is unlikely to be sufficient.

Ventilation and heat recovery

The average building is more likely to benefit from extractor fans running intermittently in individual rooms (particularly bathrooms and kitchens) ideally in association with a background whole-house trickle ventilator which runs continuously. Whole house ventilators can also be fitted to heat recovery units, but this investment will only save money if the property can pass a blower test with < 5 $m^2/h/m^2$ or better. If they have an air-quality filter fitted, remember this has a small associated energy cost and is best kept as clean as possible. The range of heat recovery units available on the market will suit any size of home or office and some models can recover up to 85 per cent of the heat in the outgoing air stream.

Heating the Home and Office

This book is not the place to discuss conventional boiler options. Chapters 3, 5 and 8 provide information on alternative heating options, specifically: wood energy, heat pumps and hydropower. However, if you are considering a new conventional boiler it is good to go for a more efficient condensing boiler. Grid-connected electricity-based heating systems are almost always the least efficient, not because of the appliance, but because the source of electricity is already reduced to between 30–40 per cent efficiency after losses in generation and distribution before arriving at your premises.

A UK survey by the Energy Efficiency Partnership for Homes concluded that the carbon impact of heating and hot water was (all else being equal) smallest in the following types of systems:

- community heating and CHP, fuelled wholly or mainly by biomass;
- wood burning boilers;
- ground source heat pumps with low temperature heat distribution (e.g. under floor);
- solar water heating combined with boiler system.

Planning to install a boiler – whether it's to provide heating or process heat for anything from a family home to a factory and industrial process – requires clarification of the main objectives. Usually, this is a combination of hot-water provision and thermal comfort in a building within the constraints of a given budget. The next task – working out how much heat is required to do this – is critical. Once the heat load has been defined and refined, then the boiler (or boilers) can be sized properly. Too big and it will produce low efficiencies and higher fuel costs. These calculations are important when it comes to installing a renewable heating system, such as a wood-pellet or log boiler (see Chapter 3).

In simple terms, the process involves the following steps:

1. List spaces for heating.
2. Work out floor area of spaces to be heated.
3. Establish ideal temperature for each space to be heated.
4. Use these figures (plus any information on building heat-loss or air-changes) to work out heat requirement for each space. The results can be used later to aid in specifying radiators.

5   Sum these for space heating load (if they are particularly tall rooms – e.g. over 3 m – then an adjustment upwards must be made).
6   Add in other load requirements for the property (e.g. domestic hot water).
7   Add the result of Step 5 and Step 6 together to get the basic heating load.
8   Take into account other considerations (e.g. other significant thermal inputs from a solar water collector or lots of lighting emitting radiant heat).
9   Add the result of Step 7 and Step 8 together to get the ideal boiler capacity.

Heat-loss calculations

For the heating system design of a new-build it's possible to apply general rules based on material specifications for the building. When it comes to older buildings it is a good idea to have a survey which includes heat-loss calculations based on a range of factors (see list below). It is always advisable to involve a professional energy auditor for a room-by-room assessment of the premises along with an analysis of past fuel and utility bills. A thorough audit could include a blower door test (see earlier in this chapter).

Given local climatic conditions, the boiler capacity is mainly determined by the amount of heat output demanded by the property. This can be calculated on a room-by-room basis, selecting radiators of specific outputs to satisfy heating requirements in the different spaces. Radiator outputs are standardised (kW or BTUs). For instance, if you have three rooms each with a 4 kW radiator and two with 3 kW radiators, then the peak space-heating requirement can be stated as 18 kW. Domestic and any other hot-water requirements also have to be calculated and added in to get to the total heat output demanded by the property. If this is estimated at a 2 kW peak, then the total heat demand should be 20 kW. This is a good indication of the size of boiler needed for the task in hand.

Key factors include:

- local climate (looking at temperature averages and extremes plus winter length);
- orientation of premises (looking particularly at building exposure);
- size, type and shape of premises (one or more storeys, detached or terraced, surface areas facing exposure issues etc.);
- insulation levels (roof, walls, floor);
- window area, locations and type;
- rates of air loss (or infiltration);
- building occupier behaviour and level (number of occupants, lifestyle pattern relating to heat and hot water issues. Something that can vary greatly e.g. between a care home and a townhouse occupied by a single person who works and plays out of home);
- occupant comfort preferences (room temperatures and use patterns);
- other heat inputs to premises (significant solar gain from a large south-facing window or unheated conservatory, solar-thermal collectors, plus types and efficiencies of lights and major home appliances since all of these give off varying amounts of heat).

In the US a commonly used reference work for domestic installations is the book *J Manual Residential Load Calculation*, published by the Air Conditioning Contractors of America (ACCA). For the UK a very useful booklet *Domestic Heating – Design Guide* is published by the Chartered Institution of Building Service Engineers (CIBSE). Wherever they are located, property owners should always ensure that contractors and/or boiler suppliers use a correct sizing calculation before signing a contract. It's a good idea to get more than one potential supplier to go through this process for you. The comparison of different suppliers' sizing results could provide important information. When the contractors are finished, always get a copy of their calculations. If you want to be better prepared before calling a supplier or contractor, then boiler-sizing services can often be accessed for a fee through most utilities and good heating contractors. It's always easier to accurately estimate boiler requirements for new-build properties where the building area and insulation levels are already known.

Predicted use patterns (user behaviour)

Essentially this will determine the room temperatures for premises, and with sophisticated controls and zoning this can vary from space to space and time to time. An average use pattern can usually be calculated by making intelligent assumptions about predicted use. DIY energy audits can usually pick up many problems in any property. Keep a checklist and inspect room-by-room, as well as outside, to help give an accurate picture and to prioritise any energy efficiency improvements that may be needed.

Suggested target temperatures per room are:

- Main living and eating areas: 20–21°C
- Office and workspaces: 18–20°C
- Kitchen and bedrooms: 15–16°C
- Bathroom: 21–23°C

Clever software packages can be obtained, often free or on a trial basis, from a variety of sources in both the US and the UK. Many of these work very well and some offer options for renewable inputs, such as wood pellets. Most are available online or as downloads from the relevant websites. In ideal circumstances, the results of a manual assessment can be checked against one or two of the available software options, prior to discussing requirements with potential boiler suppliers.

Heat distribution systems

Heat distribution systems are required for most renewable energy heating devices and, these days, intelligent modulating controls are the clever heart and nervous system of any boiler and are of paramount importance for determining overall system efficiency. If the boiler can modulate smoothly to suit user hot-water and heat demand, this will save on fuel and maximise the heat output from that used.

There are a few configuration options to consider.

Buffer tank utilisation: this helps reduce modulation and cycling in the boiler, and to provide instant hot water which can be mixed to the right temperatures for heating or hot water demands, a buffer or accumulation tank is a sensible option for most larger buildings, like office blocks and workshops or factories with specific hot water needs.

Solar thermal input: similarly, if the property has a good solar aspect, it is worth considering investment in an additional thermal input in the form of an array of solar panels which can be plumbed directly into the buffer tank (or special solar cylinder) so that it can share its heat with the rest of the system. See Chapter 4 for details on this technology.

Standard distribution systems: the two main wet heat-distribution systems used by wood pellet boilers are radiators and underfloor systems. A relatively

---

Manual method for calculating a building's heat load

The worked example below is an approximate attempt to assess boiler size for a detached rural house in the temperate weather zone. There are no extreme weather conditions and the temperatures rarely go below -3°C. Preferred internal temperatures are 21°C.

This is a two-storey house 12 m long by 8 m wide, giving a floor area of 96 m^2. The wall area is 200 m^2 (12 m × 5 m high to start of roof space × 2 walls – 120 m^2, **plus** 8 m × 5 m × 2 walls = 80m^2).

There are 16 windows with a total surface area of 40 m^2: 8 downstairs (each measuring 2 m × 1.8 m) and another 8 upstairs (each measuring 2 m × 1 m). In this house, as in most cases, the roof (or ceiling) area (96 m^2) is the same as the floor area.

For this example we have selected typical U-values for the floors, walls, ceiling roof space and windows. For regions where R-values are more commonly used, please refer to the box on R- and U-values earlier in this chapter and for further information there are some reasonably useful websites:

www.wers.net/FAQs
www.theyellowhouse.org.uk/themes/insula.html#in15
www.nfrc.org/documents/U-Factor.pdf
www.ecowho.com/articles/22/What_is_R-Value,_U-Value_and_how_do_they_relate_to_insulation?.html

The chart below uses U-values for its heat-load calculation. Similarly, the ventilation factor selected is fairly typical of a standard temperate-zone house. By working through this chart making manual calculations, it is simple enough to arrive at an approximate heat-loss figure for the property. Then, by adding an estimate for domestic hot-water demand (we have assumed 1000 W per person living in the property) the resulting figure is the required boiler size.

To calculate the annual heating demand, simply multiply the kW boiler size by the peak load equivalent hours (which is generally estimated as 1,200 hours for UK climatic conditions): 24 × 1200 = 28,800 kWh.

To calculate the annual fuel requirement, divide the annual heating demand by the rated efficiency of the selected boiler (in this case we assume a boiler efficiency of around 90 per cent): 28,800 divided by 90 = 32,000. Then divide this figure by the calorific value of the fuel to obtain the annual fuel requirement.

A worked example:

Table 2.3 Basic heat loss and boiler sizing form for manual calculation

		A	B U-value (except row 5 which is ventilation factor*)	C Temperature differential (between expected outside lowest and preferred inside temperatures)	D Heat loss for property W (Watts)	
1	Floor area (m²)	96	1	24	2,304	Multiply floor area (1A) by U-value (1B) then multiply the result by the temperature differential to get heat loss of floors (here we assume −3°C outside and 21°C internal; preferred temperatures)
2	Wall surface area (m²)	200	1.68	24	8,064	Multiply wall surface area (2A) by the U-value (2B) then multiply the result by the temperature differential to get the heat loss from the walls
3	Windows surface area (m²)	40	2.8	24	2,688	Multiply window surface area (3A) by the U-value (3B) then multiply the result by the temperature differential to get the heat loss from the windows
4	Roof area (ceiling area m²)	96	0.6	24	1,382	Multiply ceiling surface area (4A) by the U-value (4B) then multiply the result by the temperature differential to get the heat loss from the windows
5	Building ventilation or air loss (5A should be the building volume m²)	480	*0.33	24	3,802	Building ventilation loss = 5A (calculated as building height to start of roof × building floor area) × 0.33 × average temperature differential
6				INITIAL BUILDING HEAT LOSS ESTIMATE	18,240	

	A	B U-value (except row 5 which is ventilation factor*)	C Temperature differential (between expected outside lowest and preferred inside temperatures)	D Heat loss for property W (Watts)	
7	Compensate for extreme weather exposure conditions (if any) by adding roughly 5 to 15 per cent			19,152	For this example of detached rural hill based house we have added 5%
8	Expected domestic hot water demand (W)			1,000	Assuming 250 W per occupant (with four people living here) = 1,000 W for domestic hot water demands
9	REQUIRED BOILER CAPACITY (W)			20,152	
	BOILER CAPACITY (kW)			20	Divide 9D by 1,000 to translate from W into kW

Table 2.4 Showing the delivery temperatures required for a range of standard heat distribution systems

Type of heat distribution system	Delivery temperature °F	Delivery temperature °C
Underfloor	86–113	30–45
Low temperature radiators	113–131	45–55
Conventional radiators	140–194	60–90
Air	86–122	30–50

Source: Energy Saving Trust, www.est.org.uk

new product on the market, Thermoplastic, offers a third option. Around twice a year it's a good idea to bleed trapped air out of any water-based distribution system. Also, make sure that radiators are not blocked by furniture like desks, bookcases or sofas:

- *Radiators*: metal radiators of one sort or another are the most common device used for wet heat-distribution. Hot water from the boiler is pumped through pipes into a radiator which, these days, are usually made from sheet steel. Inside the radiator, the hot water circulates and gives out much of its heat to its metal fins or plates (usually steel) designed to increase the surface area of the apparatus and transfer heat the surrounding air.
- *Underfloor*: underfloor coils are an increasingly common option for wet heat-distribution. Suited principally to ground-floor areas, and most ideal in a new-build scenario, a circuit of coiled pipes can be laid under a floor. Often laid on top of sand under a concrete or tiled floor, the heat rises from the fabric of the floor itself into the room. It can be very comfortable to live with, though some people complain that it makes the air too dry. The hot water in under-floor systems usually runs at much lower temperatures than in radiators. Another advantage of underfloor heating is that it takes up absolutely no wall space.
- *ThermaSkirt*: designed and manufactured in the UK by The DiscreteHeat Company Limited, ThermaSkirt is a product that replaces radiators and skirting boards in one. Like underfloor heating it heats a room from all sides and at low temperatures. It is wholly compatible with wood pellet boiler systems and claims 13 per cent more efficiency than radiators. Also, unlike radiators, ThermaSkirt allows for furniture space along all walls. ThermaSkirt uses a unique aluminium alloy material that is said to be five times more effective than steel at passing heat and can run efficiently at temperatures between 35°C and 75°C.

Boiler controls: most modern central heating systems – like oil, gas or good wood pellet boilers – incorporate, as standard, sophisticated system and appliance controls. Today, these sometimes include external broadband links to the programmer and internal wireless connection to room thermostats. Once up and running, a boiler can be switched on and off according to demand by either a room thermostat or the hot water cylinder thermostat. The boiler itself has an off/off switch and an ignition device. There will be an electronic programmer which can control boiler functions by applying operator programmable stop and start times for the various functions (primarily space heating, hot water provision).

These days most central heating systems include individual room thermostats, making it possible for users to enjoy different temperatures in different spaces. Generally speaking, a room where people sit and do little activity may benefit from a higher temperature than a bedroom or workshop space. Radiators themselves are also commonly fitted with control valves which allow the user to give priority to radiators in certain rooms over others in the system.

There are a few basic rules of thumb for managing heating controls:

- Make sure thermostats are well located (away from draughts or heating source).
- Use thermostatic radiator valves to control individual space temperatures.
- Try to avoid a free for all with the temperature control (make one person in the home or business responsible for setting it at the right temperatures).

- Use seven-day (weekly) timers so that the weekend timings can be different (as needed by the home or business building use patterns).
- For optimum efficiency in the workplace, install weather compensation and optimum start-up controls (note that building regulations in the UK demand optimum start controls for boilers over 50 kW). Weather compensation feeds data from external and internal temperature sensors to boiler controls, maintaining internal temperatures as required by intelligently predicting demand as outside temperatures changes. Optimum start-up controls the time when heating starts in a building to ensure that when it opens for work the internal temperatures are those required (it avoids wasting unnecessary heat by responding intelligently to the early morning external temperatures and weather conditions, starting at just the right time on each day to provide precise comfort levels inside).
- If operating a complete heating, ventilation and air-conditioning system, make sure that it operates in an integrated manner (interlocked controls).
- Ensure that all systems are operating effectively and are always well maintained (check condensers regularly and clean fans, filters and air ducts at least twice a year).

Valves are needed in the system to help divert the hot water flow by giving priority to either the space heating by the radiators or the hot water requirements of the cylinder and taps. Both the room and cylinder thermostats will let the boiler, and an electronically operated diverter (or mid-position) valve, know when they need more hot water to do their job. The valve will switch the flow from one to the other depending on demand, but if both the radiators and the cylinder need hot water simultaneously, then it will automatically assume a mid position, allowing both flows at the same time.

A central-heating pump over-run feature is common in modern boilers. This makes the pump keep running and moves water out of the boiler and around the radiators and/or cylinder for a little while after the boiler combustion switches off. This protection device gives longer life to the heat exchanger by avoiding any static water remaining in the boiler and potentially boiling any residual heat in the minutes following boiler switch off.

Don't forget, too, that all hot water pipes should be insulated with foam sleeves and there should be a thermally efficient hot water cylinder jacket to ensure basic levels of energy efficiency.

3
Wood Log and Wood Pellet Heating

Burning wood to keep warm has a very long history and we have used it for processes like cooking and pottery-making for more than 10,000 years. Until just over 200 years ago, prior to the fossil fuel age and the industrial revolution, virtually the entire human race depended on wood as their main source of heat. In recent years combustion technology has advanced significantly on the basic wood-fired underfloor heating used by Vikings, Anglo-Saxons or Romans. Now, with rising oil prices and concerns about global warming, the practice of extracting energy from wood is experiencing a resurgence and a period of intense innovation in the early twenty-first century as we shift away from dependency on the diminishing supply of fossil fuels and look towards a low-carbon future.

Appliances for burning wood as logs or in the form of wood chips, briquettes or pellets are now widely available. For the average home and office, however, the range of wood combustion units is more limited due to the typically small heat-loads and problems encountered particularly with automated feed-mechanisms for wood chips, and briquettes at this small scale. Logs can be used at a domestic and small-office or workshop scale but generally require human handling, even for the loading of state-of-the-art automated log boilers.

Modern Biomass Heating: Overview

While some of us may have been brought up with open or glass-fronted coal fires as the main source of household heating, the last 40 years have seen most European and American homes convert to oil or gas-fired central-heating boiler systems, occasionally backed up with an open log fire or wood stove in the main living-room. With open fires, most of the heat goes up the chimney. They are generally rated at between 10 and 20 per cent efficient at delivering the potential heat from quality fire logs. Metal fireplace inserts are available and good ones can improve efficiencies up to around 60 to 70 per cent, but they are generally as expensive as wood stoves, which offer greater efficiencies (of percentages up to around 85 per cent or even slightly more). These days, however, a fairly wide range of wood-burning appliances are used by households, particularly in rural areas of countries with traditional forest cultures, such as Austria, Finland and Sweden. Since wood-fuel was classified as a carbon-neutral fuel under the Kyoto Protocol, most countries that require winter space-heating have begun to look

Figure 3.1a A modern wood stove with heat resistant glass door

Source: Blaze King Industries Inc., www.blazeking.com

Figure 3.1b Wood-pellet stove with back-boiler

Source: Andrew Warren

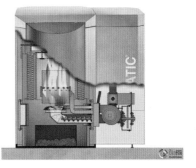

Figure 3.1c Cutaway showing combustion process in wood-pellet boiler

Source: Okofen, www.pelletsheizung.at

Figure 3.1d A Stanley solid-fuel cooking range with back boiler (large fuel box designed to 17 kW with wood logs)

Source: Dilwyn Jenkins

Figure 3.1e A 540 kW KOB biomass boiler

Source: KOB, www.kob.cc

again at wood. Indeed, a whole new manufacturing sector – well beyond the frontiers of Canada, Sweden and Austria – has developed to provide domestic scale and public or commercial end-users with stoves, cooking ranges, cooker-boilers, ceramic stoves and boilers fuelled by wood logs, wood chips or wood pellets.

Recent advances in boiler designs and controls mean that most modern boilers are highly automated. With a water-based heat distribution system already in place, it's relatively simple for any household or small business to remove an existing stove or boiler and replace it with a wood-fuelled one that will do much the same job. There are, however, a number of potential obstacles which need to be considered. Some of these are structural issues that relate to the specifics of the building that requires heating, while behavioural choices, availability of equipment and technical support as well as costs are also relevant factors. If any of the

WOOD LOG AND WOOD PELLET HEATING 53

Figures 3.2a and 3.2b A well-constructed log store with slatted sides for airflow and a decent roof to keep the rain off

Source: Dilwyn Jenkins

Figure 3.2c Not quite as well constructed a log store, it has a roof but not enough airflow since the side walls are filled

Source: Dilwyn Jenkins

Figure 3.2d A rough-and-ready wood store, it has a roof which might blow off in high winds causing safety risks and the side is too open allowing any heavy rain to dampen the logs

Source: Dilwyn Jenkins

issues pose an obstacle to using wood for heat, there may be a solution, but the cost implications of this would require careful assessment on a case-by-case basis. One example might be the need to build a new fuel store.

Wood Log Stoves and Boilers

As long as they are locally and sustainably sourced, dry and well seasoned, logs are arguably the greenest of all wood fuels. Apart from drying, they require less processing (and so less energy input, at least at the processing stage) than chips or pellets and it's even possible to gather them by hand yourself if you happen to own or have access to local woodland. As well as the quality of the fuel itself, a heating system's emissions are strongly determined by the efficiency of combustion and heat distribution systems. Fortunately, there are plenty of excellent wood-stove manufacturers all around the world. It's harder, but still possible, to find log boilers and log cooking-ranges or log-fuelled cookers with back boilers that can run wet radiator heat-distribution systems.

Wood stoves are generally made from either cast iron or rolled steel. Some are also lined with fire bricks, which make the outer surface of the stoves less dangerously hot and offer longer life to the stove itself. There are three main categories of wood stove:

- *Simple combustion*: although based on radiating heat, like radiant stoves, these are essentially simple metal boxes with a front door which can be open during combustion. With the door open, however, efficiencies are similar to that of an open fireplace.
- *Radiant heaters*: these stoves emit their heat by radiating it from the metal surface. Some have air-flow controls which can be automated and operated by a fan. They tend to have efficiencies of between 70 and 80 per cent and are probably the most common.
- *Circulating heaters*: circulating wood stoves are twin-walled which allows air from the room to flow in a circulating fashion (often with the help of an electric fan) between the inner very hot surface and the outer quite hot metal. The warmed circulating air is returned to the room as space heating and there is some radiant effect from the outer casing, too. Internally, these stoves normally have firebricks. The outer casing is usually made from a lighter-weight metal than the inner. Efficiencies are in the 70 to 80 per cent range.

Despite clear environmental advantages to using wood fuel, the utmost care is required to minimise unnecessary emissions. While this may be second nature to Scandinavians who still maintain a forest tradition and have an intimate relationship with wood, most of Western Europe and to a lesser extent North America have lost this tradition. In Scandinavia, households have meticulous systems for stacking, seasoning and drying their wood fuel and will be working two or three years ahead of themselves in order to ensure the optimum quality of logs. This process ensures maximising the heat released from the wood fuel since hardly any energy has to be used to dispel moisture. Quality wood fuel, burnt properly, will also release significantly less harmful emissions into the atmosphere and is less likely to cause carbon monoxide build up in the home or workplace.

These few simple steps will ensure you get the most energy and cleanest combustion from your logs:

- Select the most efficient stove or boiler you can afford (if in a Smoke Control area, select from the list of exempted appliances).
- Use a competent person to install the combustion and flue equipment (in the UK, solid fuel fireplace installers must either be HETAS registered or registered with the Institute of Heating & Plumbing; solid fuel burns at higher temperatures than gas or oil so flues for wood stoves, fireplaces or wood boilers are of a higher standard).
- Only use well-seasoned, dry wood (normally this will involve establishing systems that rotate wet wood with dryer wood and then dry wood in different stacks or stores).
- Have an ample out-of-house storage system to hold, if possible, at least two years' worth of wood fuel requirement (ideally the storage will allow fresh air, some wind and sunlight to enter but restrict access to rain or other moisture);
- Avoid burning plastics, polystyrene, particulate-board, plywood or wood that has been painted or treated.
- Ensure regular (twice a year) inspection of flue and chimney system for removal of soot and creosote build up.

There are a few other factors and options which should be considered:

- durability of the stove;
- use of catalytic converters;
- secondary burning/combustion option (which facilitates recirculation for burning of gases that might otherwise disappear up the chimney); and
- pre-heated air intake tubes (a simple efficiency aid).

Masonry Stoves

A single, centrally located, ceramic or masonry stove could effectively heat a very well insulated house or small office-building. The high thermal mass these types of stove store their heat in gives them some of the highest efficiency-rates of any stove. Common in Alpine countries, particularly Switzerland, ceramic stoves are often aesthetically pleasing and quite expensive. Masonry stoves are found more in North America and are generally hand-built, if not homemade. Both ceramic and masonry stoves are designed to burn just a few well-dried and seasoned logs once a day at very high gas flue temperatures.

Wood Pellet Heating

Wood pellets are a robust new fuel which is a highly versatile wood-based product designed specifically to compete with fossil fuels on convenience and performance. These days, pellets also compete well on price. Furthermore, within the wood-fuel arena, pellets tend to function better than chips in smaller sized heating-systems, paving the way for micro-level biomass. In general pellets offer a number of distinct advantages:

Figure 3.3 Masonry Heater, by Brian Klipfel of Fireworks Masonry

Source: The Masonry Heater Association, www.mha-net.org

- drier than comparable fuels such as wood chips;
- free flowing, almost fluid in nature (homogeneous in size and shape, usually around 6 to 10 mm wide and between 2 and 4 cm long);
- lower in ash content and unproblematic in ash behaviour;
- much denser than comparable fuels;
- easier to store per kWh of heat output;
- cheaper to transport per kW of heat output;
- easy to ignite (pellets are dryer and most pellet units boast push button electric starting mechanisms);
- produce lower emissions than other wood fuel applications;
- size and shape allow for optimum airflow, which is important for maximising heat outputs;
- small size facilitates loading via automated boiler feed mechanisms;
- small size allows for small-scale, highly efficient combustion; and
- cheaper than oil (and around the same price as natural gas).

The burgeoning twenty-first-century wood pellet industry is led by Sweden, the US and Canada, the latter alone producing round 1.3 million tons of pellet fuel in 2008. Pellets were first invented in Sweden to provide a flexible fuel out of waste sawdust, the co-product of a massive forestry and timber industry. They provide a simple solution to two main problems:

- What to do with sawdust mountains that otherwise accumulate at saw mills and joinery yards; and
- How to provide the world with a wood fuel that could penetrate the potentially massive small-scale biomass market.

Recognised as a virtually carbon-neutral fuel, wood pellets offer nations, companies, governments and householders the opportunity to reduce their carbon emissions more or less overnight. This factor, combined with the rising relative cost of heating oil and the reliability and efficiency of pellets at the small scale, have led to a rapid expansion in the number of pellet users. In recent years, wood pellets have also been chosen as a preferred fuel for large-scale heating and even combined heat and power (CHP) applications. They are cheaper to transport, being denser than wood chips, and provide a more consistent and drier fuel for combustion.

Households and small businesses in some European countries have switched rapidly from oil to pellets. By the end of 2008, over 62,000 pellet central-heating systems had been installed in Austria with a nominal power output of around 1,200 MW thermal. A study by the Salzburger Institute for Urbanisation and Housing indicates that an average Austrian household switching from an oil to a pellet heating system would save up to ten tons of CO_2 emissions every year.

Wood Pellet Manufacture

Pellets are made from clean wood-waste or clean forest-sources such as saw mills, joinery factories and well-managed woodland thinning. In 2007, Europe produced around nine million tons of pellets, led by Sweden and Austria. Any untreated wood waste can be used, from sawdust and wood-chip, to larger off-cuts, whole trees and stripped bark. The feedstock must be from solid wood, never from composites like chipboard.

Where freshly harvested whole trees are used much more energy must be consumed during the drying process. This has economic, energy and environmental costs associated with it. Recycled wood can also be used, but not if it has been treated. The use of treated wood is limited by national regulations stipulating the amounts of particular compounds, heavy metals and additives that can be legally present in the pellets. The UK Department of Trade and Industry's 'Codes of Good Practice' state that pellets destined for large commercial use (mainly co-firing in power stations) can contain up to 15 per cent MOB (material other than biomass) while pellets destined for domestic use must be manufactured from virgin and clean wood only.

Some form of starch or carbohydrate is commonly added to assist in binding the pellets at the pressing stage. The addition of vegetable oil can also be used to aid lubrication.

Pellets and Emissions

As with all biomass heat provision, wood pellets are responsible for CO_2 emissions at the point of combustion. When burnt, levels of pellet emissions are similar to that of coal and almost double that of natural gas. Stove or boiler efficiency also affects overall efficiencies and emissions, so it's important to burn pellets, as with other wood fuels, as efficiently as possible, particularly during the next 50 years while global society struggles to mitigate and minimise the impact of our changing climate.

Pellets and Job Creation

Pellets also boast high levels of job creation and the tendency to keep income circulating in local economies. Jobs are created in forestry and woodland management, transport, pelletisation, combustion equipment manufacture and sales. Since the mid 1990s, Austria has developed a pellet stove and boiler manufacturing industry worth over 100 million Euros a year. In 2005 alone, Austrian pellet-equipment manufacturers supplied 8,874 wood pellet boilers and 3,780 wood pellet heaters.

Automation and Convenience

One area where jobs are not created by wood pellets is in the field of loading and operating a stove or boiler. Some pellet space-heating systems are so well automated that they need no human operator or maintenance beyond loading the winter store, pressing a button to start the system, then de-ashing six months later. This level of convenience competes well with that offered by oil and gas boiler systems.

Security of Energy Supply

Wood pellets also offer a region like Europe the opportunity to improve the security of energy or fuel supply. This is true at a regional and national level, where wood-energy crops can be incentivised for heat and electricity generation, but it is also true at the level of a household or small business. Being dependent on imported oil or gas is becoming an increasing risk, as well as increasingly costly. Having access to locally grown and produced wood – whether logs, chips or pellets – is a sensible strategic move.

The supply of wood pellets to the end-user is already established in many regions of Europe and North America, but it is certainly something to check, before making the decision to go for this fuel and appliances that require it. Handy 10 kg bags can be picked up at some corner stores and petrol stations, but for better value pellets are best bought in bulk for home or office delivery. The present cost of a ten-ton delivery in the UK is around £200, within an 80 mile radius of the pelletising plant. Most wood pellet deliveries have the facility to blow pellets down a tube from the lorry, straight into a pellet store.

Structural Issues Specific to Buildings

The Need for a Fuel Storage Area

Unlike natural gas, which can be piped into every house in a city, wood fuel is delivered and has to be stored much as coal used to be. Heating oil is commonly used in rural areas and this too requires a storage tank. Wood logs and wood pellets tend to require more space per kW of heat output than oil, so where the fuel is to be stored is an important consideration. Wood pellets, in particular, need careful storage and handling, especially if they are to be automatically fed into the boiler. This is particularly true for boiler systems or stoves that provide a building's main heat source (i.e. installations that will get through a few tons of fuel during a winter). The usual options are sheds, old coal stores or purpose-built wood or pellet stores.

Fuel Handling Requirements

This really relates to both the delivery of wood fuel to a heating installation's storage zone and also delivering the fuel from store to combustion unit. If the installation requires the manual handling of wood logs, proximity or access with a wheelbarrow are typical considerations. A wood-pellet installation, however, has more complicated spatial needs. Pellets are small and quality pellets arrive almost dust-free. In small amounts they are handled in 10 kg bags, but most pellet uses buy in quantities of 1 ton bags which need dry storage, ideally not too far from the boiler room.

Figure 3.4 Wood pellets are clean and dust-free to handle

Source: Dilwyn Jenkins

Boiler Footprint

Both stoves and boilers have footprints (i.e. the space they physically inhabit plus any space needed around them for loading, operating and cleaning the appliance). This needs to be thought through and ideally mapped in relation to specific heating appliances being considered for installation.

Figure 3.5 Wood pellet boiler and footprint

Source: Dilwyn Jenkins and Okofen, www.pelletsheizung.at

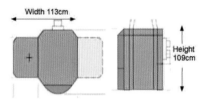

Fuel Availability and Price

Wood-fuel prices have varied less than oil or gas equivalents in the last few years. This process is likely to continue to improve with logs, chips and pellets becoming relatively cheaper as fossil fuels become scarcer and more expensive. These days, outside of rural areas, it's arguably easier to guarantee availability of wood pellets than quality wood logs. But whichever of these is the chosen fuel, it's important to know in advance where you'll be sourcing it from. Generally speaking, logs are considered too expensive if they travel more than 50 miles, while pellets are often distributed economically within a wider radius (80 to 250 miles). Similarly, before making the final decision on the purchase of a wood-combustion unit, it makes sense to get a good feel for the price of the chosen fuel and, as far a possible, some idea of how this price is likely to evolve in the longer term.

Spatial Impacts of Available Automated Fuel-Feeding Mechanisms

Since we are not including wood-chip boilers as they are not appropriate for the average home or small-business heating application, this point refers mainly to wood-pellet automated fuel-feed mechanisms which tend to be either vacuum suction- or screw-auger-based systems. Each of these has their own implications, particularly in relation to building layout and proximity of fuel store. There is more on this in the section on wood pellets.

 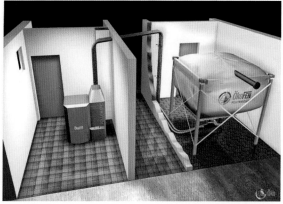

Figure 3.6 Wood pellet automated fuel-feed mechanisms: (a) screw auger; and (b) vacuum operated

Source: Okofen, www.pelletsheizung.at

Generalised Issues Relevant to Wood Heating Installations

Relative Costs of Appliances

For most householders and small businesses, the available budget is always a limiting factor when it comes to capital investments like new heating-systems. Given that the range of stove and boiler heating-system costs being considered in this chapter runs from around £500 (around US$818) to a possible £20,000 (about US$32,700), a close analysis of costs (including costs of delivery, installation and commissioning, and any post-sales services, common replacement parts, etc.) is a good starting point.

Sizing Combustion Equipment Correctly

Wood-fuel combustion units should be selected to meet the precise heat demands of the property in question. This ensures the best combustion efficiency and a smoother delivery of the required heat and hot water. Oversized combustion units will cost more in terms of capital and ongoing fuel needs and, if associated with automated wet heat-distribution systems (see Chapter 2), will experience additional inefficiencies from cycling that is too frequent (essentially the stopping and starting of combustion process). Undersized combustion units will fail to provide the desired levels of space heating and/or hot water.

In simple terms, the process involves the following steps:

1   List spaces for heating.
2   Work out floor area of spaces to be heated.
3   Establish ideal temperature for each space to be heated.
4   Use these figures (plus any information on building heat loss or air changes) to work out heat requirement for each space. The results can be used later to aid in specifying radiators.
5   Sum these for space heating load (if they are particularly tall rooms – e.g. over 3 m – then an adjustment upwards must be made).

6   Add in other load requirements for the property (e.g. domestic hot water).
7   Add the result of Step 5 and Step 6 together to get the basic heating load.
8   Take into account other considerations (e.g. other significant thermal inputs from a solar water collector, expected simultaneity in heat demands, or lots of lighting emitting radiant heat as well as dynamic effects).
9   Add the result of Step 7 and Step 8 together to get the ideal boiler capacity.

For a property owner and potential wood heating boiler-user, going through this stepped process will provide a working notion of the required boiler capacity, but it is always very important to get your boiler manufacturer, authorised dealer and/or qualified installer to go through the process according to their own techniques and methodology. After all, they are the ones offering a warrantee and if they are any good they will already have lots of experience at sizing pellet systems.

Rules of Thumb

Although useful, particularly when gained through the experience of designing and installing a specific boiler technology, 'rules of thumb' are not enough on their own. They can be helpful in the initial stages, in preparation for early discussions with boiler manufacturers, suppliers or installers.

Based around a building's total footprint (ground-floor area), the boiler output should be 1 to 1.5 kW per 10 m^2 of floor space (this is very crude and assumes good insulation levels, average ceiling heights plus temperate weather conditions). Assuming a ground-floor area of 96 m^2, the formula reveals a boiler-size estimate of between 10 kW and 15 kW.

Based around building volume (ground floor area only $\times$ height to start of roof space, e.g. 96 $m^2 \times$ 5 m = 480m^2), multiply volume (e.g. 480m^2) $\times$ 40 = 19,200 W or 19 kW. In this example, a 19 kW boiler might be specified, though this rule of thumb is based on very low insulation levels and would almost certainly be downsized by up to 25 per cent, depending on insulation levels and air loss factors. In this way, the boiler specified might be in the range 14 to 17 kW.

A stove rated at 60,000 British Thermal Units (Btu) can heat a 2,000 square foot home, while a stove rated at 42,000 Btu can heat a 1,300 square foot space (US Department of Energy).

Emissions Regulations in your Region

Emissions regulations and the way they work vary significantly between Europe and the Americas. In the US the installation process is generally less regulated. There is some national regulation from the Department of Housing and Urban Development (HUD), but most aspects are covered by local building codes. Generally these relate to the installation itself, rather than who carries it out. In mainland Europe, regulation again focuses on the installation. Germany and particularly Austria have high emission-standards to be met by manufactures. Also, many countries require an annual service and chimney clean by a qualified professional who provides the householder with a certificate. The relatively small-scale combustion units needed by most homes and small businesses (up to 50 kW thermal) have to meet fewer regulations than larger boilers, but there are still considerations in relation to emissions and building control. In the UK, for instance, most urban areas are 'smoke control' zones where, in order to burn wood fuel a property would need to install one of a growing list of exempted heating appliances which are specific makes and models.

Availability of Equipment (Installation and Post-sales Support)

Much less of an issue than it was even a few years ago, the availability of combustion appliances – plus the skills to install it and provide ongoing maintenance support – is still a factor that needs investigation at the local level. You can approach this matter from the top down or bottom up (i.e. by identifying the technology first and then investigating local suppliers next, or by talking first to a reputable stove and/or boiler supplier in your region and taking their advice on makes and models).

Availability of Advice and Financial Support

This is probably the least important issue to bear in mind, but still potentially very useful. Advice might come from your regional Energy Agency or Energy Advice Centre, from a local supplier, at an exhibition or increasingly online from a number of possible sources. Financial support will depend on government or local-authority grants (see Annex 1 for information on finance and support). When and where capital grants are available to domestic or small-business property owners or operators, they are typically between 25 and 50 per cent of capital costs: a substantial help.

Level of Automation

Figure 3.7 Fully automated wood-pellet boilers make them a strong contender for home, office or workshop situations – as this picture shows, even the delivery of pellets can be automated

Source: Balcas Ltd, www.brites.eu

High Combustion Efficiencies

Pellets stoves and boilers tend to offer high levels of combustion efficiency, even at the small end of the heating scale.

On-going Fuel Costs

In theory, but not always in practice, logs should be the cheapest wood fuel. This is clearly true if you can source it yourself or from reliable local suppliers. A common problem with logs is that they are often not fully seasoned or sufficiently dried. Pellets are made to high standards and can compete with oil and gas on convenience because of their fluid nature, dryness and density in relation to heat output. Even more attractively, they also represent a cheaper fuel option, at the moment, than heating oil.

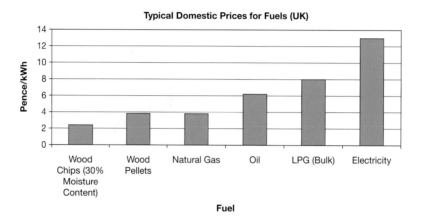

Figure 3.8 Fuel Price Comparison UK, late 2010

Source: Biomass Energy Centre, www.biomassenergycentre.org.uk

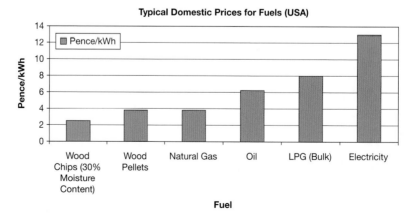

Figure 3.9 Fuel price comparison USA, 2011

Source: Pellet Fuels Institute, www.pelletheat.org

Behavioural Factors

Behavioural factors here refer on the one hand to the limits on human inputs associated with the heating process and on the other to the nature or pattern of demand for that heat. Wood logs, for instance, usually require stacking by hand and moving physically from the store to the stove or boiler. This may not suit everyone in terms of available time or physical ability. Logs also need to be loaded in a timely manner, either when needed to keep combustion going in a wood stove or stacked for a day's combustion as in some log boilers. The timing of these events may not suit the lifestyle or work style of a building's occupants, for instance it may not be possible to keep an eye on the stove to make sure it's fed when necessary.

The pattern of use of homes, offices, workshops and community spaces can pose difficulties for implementing wood heating. Community centres or church halls are a classic example, some being used as little as three mornings a week, just for two or three hours at a time. Of course, it's possible to heat for these occasions with a wood stove or boiler, but it would require significantly more human input than an electric heating system, an instant appliance that provides heat at the flick of a switch. There will be more on this in the section on Community Heating, below.

De-ashing Requirements

De-ashing a wood stove or boiler is a task avoided with gas, electric and oil heating-systems. Some of the smarter wood pellet boilers have ash compressors and only require emptying once or twice a year. More basic stoves, wood ranges and boilers generally require de-ashing at least once a day. The range of possibilities is large and might well be a factor that some wood heat users would like to consider while selecting a combustion appliance. This is the kind of information that most manufacturers and suppliers of stoves and boilers will be happy to advise on.

Figure 3.10 Wood stoves need regular de-ashing, but pellet boilers tend to keep this to a minimum, sometimes requiring it as little as twice a year. This ash chamber has a compacting device.

Source: Dilwyn Jenkins

Energy Efficiency

With all renewable sources of energy it is important to incorporate energy efficiency into every stage of a system, from fuel delivery, combustion and heat-distribution design right through to the fabric of a building or appliance. Energy efficient heating systems in energy efficient buildings clearly keep annual fuel costs to a minimum. The best place to start with advice on, and grants for, insulating homes and offices is your local authority or nearest Energy Agency or Energy Advice Centre (see Annex 2).

All good wood log and pellet stoves and boiler manufacturers provide specifications with combustion efficiency. Many of the best now boast efficiencies between 85 and 95 per cent. The combustion efficiency of wood stoves and boilers is also very dependent on the quality of the wood fuel itself. In the case of wood logs, this can be highly variable. Pellets in Europe, on the other hand, have to maintain standards of low moisture-content in order to be sold as wood pellets for heating purposes.

In the US, Pellet Fuels Institute (PFI) members have developed standards for residential and commercial densified fuel and initiated the redevelopment of its standards in 2005. They are also planning to implement a program that, it has been proposed, will be incorporated by reference into the US Environmental Protection Agency (EPA). The PFI Standards Program hopes to ensure that fuel pellets will be certified to a specified grade and can be properly matched to the appliances that are permitted to burn them. The main established standards in the EU are the DIN 51731, plus the Austrian ÖNORM M1735 and the Swedish SS 187120 and SS 187121. The EU-wide CEN standard EN 335 for solid biofuels (including wood pellets) is presently in preparation by the CEN/TC 335 technical committee. They are developing the draft standard for solid biofuels within Europe, including wood chips, wood pellets and logs.

Regulations and Emissions Control for Wood Heating Installations

In the US, the EPA provides standards for air quality, covering a number of principal pollutants, which are called 'criteria' pollutants: carbon monoxide, lead, nitrogen dioxide, particulate matter and sulphur dioxide. The US Clean Air Act, through the EPA standards, requires that wood stoves and wood heating appliances meet a particulate emissions standard of no more than 7.5 g per hour for non-catalytic wood stoves and 4.1 g per hour for catalytic wood stoves. In the UK, any heating installation between 400 kW and 3 MW is required by the local authority to have an air-pollution control permit. Heating systems any larger than 3 MW are regulated by the Environment Agency, which is responsible for implementing the Pollution Prevention and Control Act (1999) and is likely to insist on emissions monitoring (either continuous or periodic).

Pollution control regulations are, of course, subject to change, particularly during times like these with increasing international concern about climate change. At the time of writing, however, smaller systems in the UK (400 kW or less) are not required to obtain any air pollution permit, although the regulations surrounding 'Smoke Control Areas' imply the need to use an exempted heating

Figure 3.11a and 3.11b (a) The Ludlow woodstove and (b) The Stretton inset stove from AGA, both of which are smoke control exempt

Source: AGA, www.stoves.agaliving.com

appliance (see below) if the building is located within one of these zones. Appliances that are registered can be used even in smoke control areas as long as the fuel does not contain halogenated organic compounds or heavy metals as a result of treatment with wood preservatives or coatings.

If you live in the UK and are in a smoke control area, your choice of combustion equipment is limited to the list of approved exempted appliances. Makes of heating equipment on this list presently incorporates many well-known brands, including but not limited to: Aga, Charnwood, Clearview, Dunsley, Ecostove, Jotul, Ökofen, Morso, Vermont and Westfire.

Codes and regulations for installation

In the US regulations vary significantly between states. In Washington, for example, the state demands that all new wood stoves draw in their combustion air directly from outside the building in order to minimise the risks of chimney backdraft, which can cause smoke and harmful emissions to enter the house. The air is usually piped in and often powered with fans.

In the UK, heating installations are covered by the Building Regulations, which are enforced by Building Control Officers. However these Officers are responsible for monitoring the complete process of building construction and renovation, and cannot be expected to understand all the detailed requirements of all the possible building services. The Government deals with this by setting up Competent Person schemes, and the scheme for solid fuel heating systems (including pellet boilers) is run by the official UK body for approving biomass and solid fuel domestic heating appliances, fuels and services, HETAS.

Members of HETAS are approved to install solid-fuel heating systems. Thus, if the system is installed by a HETAS-registered installer the system is considered to meet the regulations, and the installer can provide a certificate to say so. Unfortunately, the rules that HETAS installers follow were developed to cover coal boilers and they are not always appropriate for wood-fuelled heating systems. However, this issue is being addressed as wood heating becomes more popular in the UK.

Most electrical work in the UK is carried out by qualified electricians, but there are no regulations to prevent a homeowner, for example, wiring in their own system. However, in theory at least, this work would then have to be approved by the Building Control Officer

Most countries have an equivalent collection of regulations. In the US the installation process is generally less regulated than in Europe. There is some national regulation from the HUD, but most aspects are covered by local building codes. Generally these relate to the installation itself, rather than who carries it out.

The installation of catalytic converters is also offered to meet clean-air requirements in urban areas in North America. Usually ceramic based with a platinum or palladium coating, catalytic converters complete the combustion of unburned hydrocarbon gases emitted by the wood fuel as it burns. This can improve a stove's efficiency by over 10 per cent and, at the same time, remove many of the potentially polluting emissions. Unfortunately they need to be replaced frequently, since they don't usually last more than three to six years.

In mainland Europe, regulation again focuses on the equipment installed. Germany and particularly Austria have high emission standards to be met by manufacturers. Also, many countries require an annual service and chimney clean by a qualified professional who provides the householder with a certificate. This emphasis on high-quality equipment and regular maintenance provides an excellent complement to the UK's focus on the detail of the installation, particularly in relation to the flue design.

Generally in Europe, the standard of an installation is regulated by the local or national building codes, which vary enormously in their ability to regulate pellet systems. However, it is possible for the client to specify that the installation must be CE marked (the mandatory mark showing a product's conformity with EC directives and a pre-condition for placing a product on the market in the European Economic Area). It then becomes the responsibility of the installer or prime contractor to ensure that the whole installation is installed and commissioned to this standard and that the necessary documentation is produced. This is different from the CE marking of individual items of equipment, such as the boiler, where it is the manufacturer's responsibility to comply.

Disadvantages with Wood Heating

Wood heating will not suit everyone. The main disadvantages are in the need for human operator input, fuel handling and mess:

- the generation of ash;
- wood particles and bark may need regular tidying;
- chimney soot and creosote build-up (which needs regular cleaning for safe operation in terms of both fire and pollution risks);
- handling of wood (often from woodland to store and then to property for heating, which is clearly avoided by using wood pellets in an automated system that incorporates a good fuel store to combustion unit delivery mechanism);
- requirements of continual loading (generally avoided by wood pellets).

Community Heating

The term 'community heating' generally applies to a shared-use community property like a village hall, or to a collection of houses or businesses that share a single heating system. In terms of community heating with wood, there is an ongoing tradition of this in Scandinavia and Austria, where wood chip or pellet district-heating systems connect everything from small clusters of buildings to whole towns.

Within the scope of this chapter, looking at heating systems of up to 50 kW output, the most likely scenario under the name of community heating is that of the community hall, though the target building could just as well be a large office or workshop space, or even a small group of such premises. These days many halls use electricity for heating because of the instantaneous effect. Halls tend to be under-utilised, fairly large and often cold and un-insulated spaces. The under-utilisation is a behavioural issue which could be addressed to make wood heating a more practical possible option. At the moment, if a community hall is only used four or five mornings a week for a few hours, instant heat is advantageous, saving on the preparatory time required for the building to heat up (as would be needed with a wood-, gas- or oil-fired radiator system).

If the hall is used for longer blocks of time (a behavioural factor, open to an element of human choice and adaptation), it is easier to justify someone arriving 40 minutes before the others to ignite the boiler of a morning. One wood fuelled response to this would be a fully automated pellet boiler with programmable timer. Like a conventional oil or gas radiator-based central heating system, this can be programmed to fire up and turn down according to the needs of any building. It is, however, a more expensive option in terms of capital investment, although it is a much greener heating solution in terms of emissions and ongoing fuel costs, which are presently cheaper from pellets than they are from electricity. With the right budget there are some very efficient pellet boilers on the market which are also fitted with fully automated feed-mechanisms and time-programmable controls. Some can even be switched on remotely via a mobile phone. Pellets are also more suitable for this kind of application since they are a smaller, drier fuel allowing the boiler to be fired up and turned down relatively fast.

Wood pellet and log boiler-systems are available for much larger heat-loads and, as mentioned above, the technology exists for several or many households or business units to share a single medium- to large-scale wood combustion unit via a district heating system: usually a wet-based system of insulated pipes to send hot water to users and return the resulting cooled water back for reheating. Heating for community buildings, schools, leisure centres and public offices is already a growing industry across Europe, at least partly as a result of financial incentives and carbon emission targets, both voluntary and statutory. At scales of 200 kWth (kilowatt thermal) or more, most wood-based boilers are designed for wood chips, which are cheaper to produce than pellets. Many of the automated wood-chip boilers are also able to burn sawdust, briquettes or even straw and grain.

Wood pellets are spearheading the micro end of the biomass revolution by offering a wood fuel that could work well with automated fuel-feeding mechanisms right down to the individual room or house level. Nevertheless, pellets are economical and also green enough to be used nowadays to fuel combined heat and power stations in Europe and North America.

How to Select the Right Heating System

Table 3.1 provides a simple assessment relating the size and type of property to the most appropriate different types of heating systems.

Once you have selected the right type of heating system, refer again to the criteria listed in this chapter to fine tune the details of the appliance to make sure they match the heating and user needs of the property. It's important to remember to check whether or not the premises is within a smoke control area. If it is, then only the approved appliances are available to select between.

Table 3.1 Wood heating systems comparison chart

Fuel Type	US$ cost per Million BTU	US$ cost per ton	US$ cost per gallon	US$ cost per kWh	US$ cost per therm
Wood pellets	18.67	245			
Oil	33.25		3.58		
Electricity	35.16			0.12c	
LPG	39.72		2.83		
Natural gas	17.38				1.39

Source: Dilwyn Jenkins, developed from an earlier version in the *Woodfuel in Wales* booklet, Powys Energy Agency 2003

4
Solar Thermal

Solar Heating for Water and Space: An Introduction

In this chapter we are going to discuss how to use solar energy for heating, firstly for water heating (the most common use), and then for other uses such as provision of space heating, or even cooling. Solar energy is electromagnetic radiation emitted by the sun. The amount of this energy that reaches the Earth is more than 15,000 times our current global energy consumption. The question is: how can we best capture it and make it useful? One-third of the sun's energy is used up in heating our air, ground and water bodies. Roughly another 30 per cent brings life to the hydrological circle, generating evaporation, rainfall and river flows. Photosynthesis for plant growth accounts for significantly less than 1 per cent of the sun's energy. The sun brings life energy to the planet's biosphere and, at the same time, offers us a permanent and renewable source of heat. Even in Northern Europe the amount of solar radiation available ranges between 600 and 1000 kWh/yr per m².

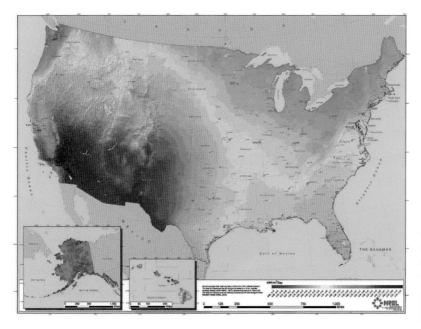

Figure 4.1a Daily annual average solar radiation on south-facing surfaces tilted at the angle of latitude in kWh/m²/day, US

Source: National Renewable Energies Laboratory (NREL), http://www.nrel.gov

Figure 4.1b Yearly sum of solar radiation on south-facing optimally inclined surfaces in kWh/m², Europe

Source: PVGIS © European Communities, 2001–2008, http://re.jrc.ec.europa.eu/pvgis/

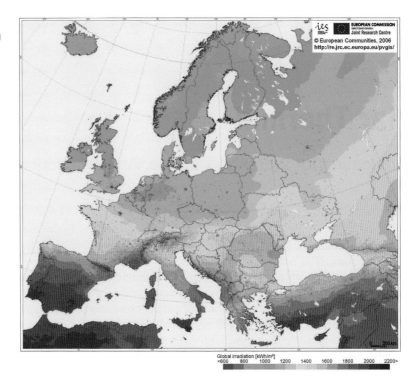

It is common knowledge that a tub of water left in direct sunlight will warm up, and that if it's painted black it will warm up more quickly. Solar water heating is a technology which uses this solar warming effect to provide hot water for domestic use or commercial processes. The simplest form is the very basic solar shower pack: available for less than US$20, they're little more than a black plastic bag with a short hose ending in a simple shower spout. The bag is left in the sun for a few hours until sufficiently warm, then hung up above person-height (perhaps on a tree branch while camping), before opening a small tap on the hose which allows the warmed water to flow down in shower mode onto a person below.

Figure 4.2 A simple form of solar shower, from a company that specialises in camping equipment

Source: Gelert Ltd, www.gelert.com

Most solar thermal systems, however, are more sophisticated and permanently sited systems involving specially designed absorbers, electric pumps and permanent plumbing into a building's heat distributions system, often, but not always, via a hot water tank. Many systems these days are roof integrated, forming part of the building's surface, adding value to properties as well as saving money compared to using electricity or other forms of heating. Domestic solar water-heating systems cost from around US$3,500 (a little cheaper if you can do much of the design, fitting and plumbing yourself) and can be installed professionally in just a day or two. They are compatible with most boiler systems and come with fully automated temperature and timer controls.

Solar pool or hot-tub heating is another very good use of solar radiation. On commercial swimming pools (e.g. in hotels) the payback period for solar heating is actually very fast. Capital costs are typically quite low since the solar absorbers may not need to be glazed and it's often possible to use the existing pool pump to circulate the heated fluid rather than have to invest in a separate one. Simple solar hot-tub heaters can even be made from improvised materials.

The economics of solar water heating have long been among the most promising in terms of payback, particularly when compared to the cost of heating water with a standard electric immersion-heater. Exactly how much money a solar thermal system will save on bills does depend on obvious factors like the amount of sunshine available at the location, the efficiency of the system selected and the hot-water use-pattern. In the right situation – particularly where there's heavy

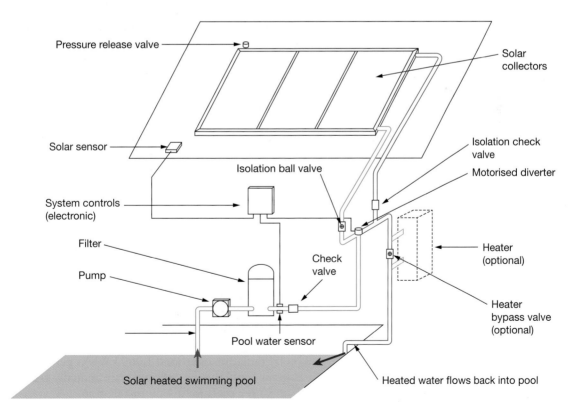

Figure 4.3 Schematic of a solar pool-heating system showing all the main components

Source: Teilo Jenkins

commercial use of hot water – payback can be expected within about five to ten years, although it is less likely to be quite so fast in a domestic scenario, unless capital grants are accessed or the system is homemade and consequently very inexpensive.

Space heating is another solar-thermal technology option which we look into in this chapter. It is expensive and systems are usually rated for the average heating load rather than to meet peak demand on the coldest days. Perhaps ironically, solar space-heating systems are often more appropriate to locations with long, cold winters.

This chapter also has a section on passive solar design, which has the potential to eliminate most of a property's heating demand via a combination of capturing solar radiation through large sun-directed glazing and then keeping the useful heat in with ample insulation and controlled ventilation. Anyone with a window on their home, office or workshop is already using a simple form of passive solar design, letting light, along with the sun's heat, into the building. To seriously tap into passive solar energy it is best in new-build or building refurbishment phases, though it is nearly always possible to improve a property's thermal properties and solar gain in an incremental fashion, perhaps by adding one or two more features and insulating a little better every year.

We also look at solar-thermal cooling, something associated more with larger buildings. Solar cooling does not have a massive uptake yet, but is presently in the development and standardisation phase. There are examples of this technology, some using heat pumps and others focusing mainly on clever desiccant or absorption techniques for cooling warm air.

The US DOE claims that water heating accounts for between 14 and 25 per cent of the energy used in most American homes. More than half of this energy can easily be harnessed by a typical 4 m^2 solar absorber system and all the normal hot water demands can be met in summer months. With fossil-fuel prices continuing to rise, the payback period on solar water heating installations is getting shorter.

Solar Water-Heating

Heating water with solar energy is the most common use of solar thermal technology. The main elements of a solar water-heating system are a collector, a pump and controller (except with thermo-syphon systems) and a dual coil solar-compatible hot water tank which acts as an energy store or buffer tank.

The Collector

Solar collectors range in type and sophistication from simple radiators painted black, to purpose made (usually metal) plates which heat water passed through them when the sun shines, or special evacuated glass tubes with higher specifications and greater hot water output. Typically, an evacuated tube design will have efficiencies of 50 per cent as compared to between 20 and 30 per cent for a simpler DIY system. Given a typical solar radiation of around 1,000 kWh/yr per m^2, this means that evacuated tube systems could theoretically generate annually around 2,000 kWh (4 × 1,000 kWh × 50 per cent) for a 4 m^2 installation.

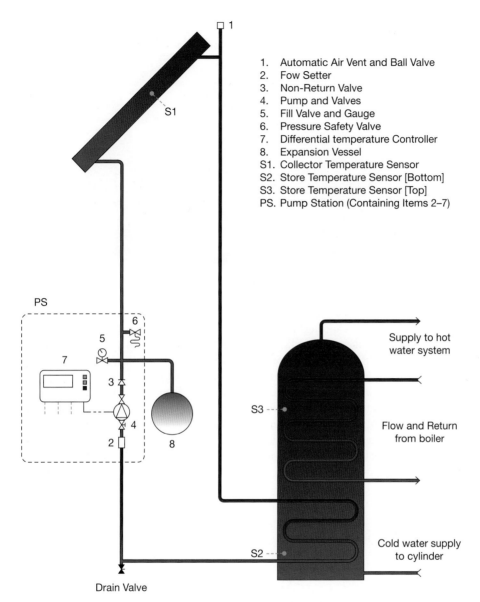

Figure 4.4 Schematic of a solar thermal system with twin coils in the hot water tank. Heat from the solar panel is pumped into the bottom of the tank when the sensor on the panel detects that it has reached the required temperature. Another sensor in the hot water tank determines whether the heat needs to be topped up by the second coil, fed from the domestic boiler. Even during the winter the solar panel will provide some pre-heating of the water in the tank – even a few degrees will give the boiler less work to do.

Source: AES Solar Systems, www.aessolar.co.uk

If the same amount of water was to be heated to the same temperatures by electricity at a rate of US$0.10 per kWh, then this solar thermal system has the potential to offset as much as $200 a year. In reality, this is likely to be somewhat less due to other inefficiencies in the system, the weather, and how hot water is used. A normal household or business is likely to function with non-optimal patterns of use for the hot water produced.

Properties with south-facing roofs (north-facing south of the equator), sloping between 20 and 50 degrees from vertical, are the most suitable for installing a solar-thermal collector, although installing onto a roof which is either south-west or south-east facing is also worthwhile. Collectors can also be installed at ground

Figure 4.5 Clip-fin collector, self-assembled. These can be purchased in kit form.

Source: Dilwyn Jenkins

level providing that there is no (or very little) shading. Collectors should not be installed onto north-facing roofs (north of the equator) or somewhere which suffers from a lot of shading. Furthermore, if there is no room in the house for a suitably sized hot water tank, the system is unlikely to produce enough hot water during its lifetime to pay for the initial financial investment.

In this chapter we focus on the two main types of solar-thermal collectors readily available in the marketplace: flat-plates and evacuated tubes. Evacuated tubes are a lot more efficient, but they are more expensive. On balance, however, they deliver similar value for money. Evacuated tubes offer the advantages of being able to produce more hot water per m^2 of absorber. This means that less roof surface is needed. They also possess the ability to heat hot water even on a freezing-cold sunny day in winter. There is a third category – integral collector-storage systems (ICS) – which range from one or more large heavy-duty rubber, plastic or metal tanks to simpler coiled or parallel black tubes or pipes. A bit like the basic solar shower, ICS tanks or pipes act as both solar absorber and hot water storage at the same time.

Flat-plate collectors

These are the lowest cost of the two main solar collector types. They consist of an absorber with a glazed layer in front creating an air gap with a layer of insulation surrounding the back and edges. Solar absorbers are made from either a single heat-conducting metal plate (usually copper because it has good thermal conduction properties) with thermal fluid pipes attached (pipe and fin) or two metal plates sandwiched together, allowing the thermal fluid to flow between them (flooded plate). The thermal fluid typically used is a non-toxic glycol alcohol-based antifreeze. In both types of flat-plate collector, cool thermal fluid flows into an inlet pipe at the bottom and heated thermal fluid flows out of an outlet pipe at the top.

Commercially manufactured collectors are typically covered in a 'selective coating' which has a high light-absorption capacity (between 85 and 95 per cent)

Figure 4.6 Typical flat-plate collector. This Sol 25 Plus Collector has a surface area of 2.7 m² (29.06 ft²).

Source: Stiebel Eltron Inc., www.stiebel-eltron-usa.com

and a low thermal emissivity (between 5 and 15 per cent). This means that when sunlight hits the selective surface most of it is converted into long-wave thermal radiation which is transferred into the metal, but that very little is emitted back out. Simple matt-black paint coatings also offer high light-absorption capacities (around 95 per cent) but they suffer from relatively high thermal emissivity (around 88 per cent), so, although 95 per cent of sunlight might be absorbed, 88 per cent of that is emitted back out.

A glazing layer allows heat to radiate through it and into the air gap between the glazing and the inner absorber surface. Not all of the solar radiation is emitted back out through the glazing, causing a greenhouse-like effect. Glazing also prevents wind from cooling down the absorber plate and guards it from moisture. One drawback to having a glazed layer is that not all of the sun's radiation is transmitted through, but the benefits mentioned above far outweigh this. The most commonly used glazing material is a highly transparent low-ferrous safety glass around 4 mm thick, giving it a maximum transmissivity (ability to let the sun's radiation through) of 91 per cent.

Insulating material behind the inner absorber surface prevents heat loss through conduction. Absorbers can reach temperatures of up to 200°C; mineral fibre insulation (such as rock wool) is placed directly in contact with the absorber, offering the necessary level of heat resistance. An extra layer of insulation with a lower U-value (making it a better insulator) such as polyurethane can be used behind the layer of mineral fibre. Polyurethane isn't as heat resistant as mineral fibre, but doesn't need to be as it is not in direct contact with the absorber.

When the sun hits the glass cover, a small part of the sun's radiation is immediately reflected back by the glass. The remainder transmits through and hits the selective coating of the absorber, which also reflects a small amount of the light. The remainder is converted into heat. The insulation surrounding the back and edges of the absorber reduces the amount of energy loss through thermal conduction and the glazed layer reduces thermal loss through radiation and convection.

Figure 4.7a Diagram showing how a flat-plate collector works

Source: Dilwyn Jenkins

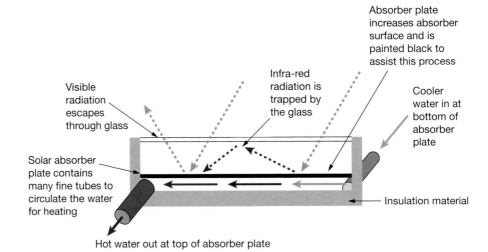

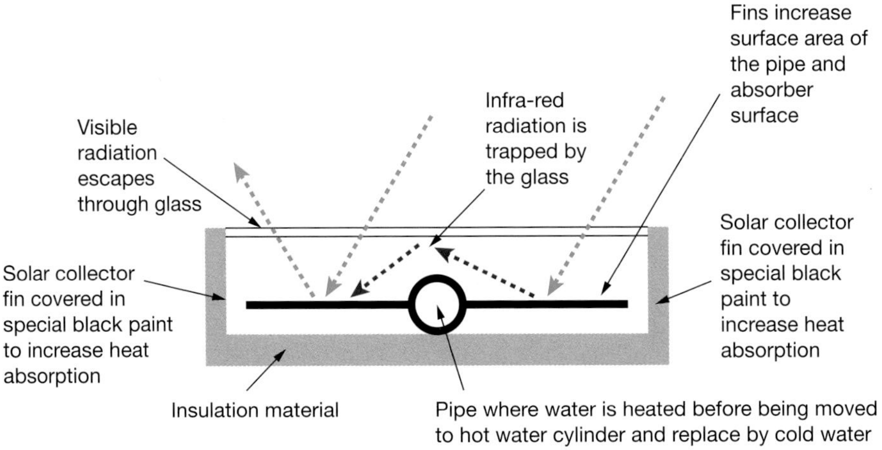

Figure 4.7b Diagram showing how a clip-fin collector works

Source: Dilwyn Jenkins

The remaining heat is conducted into the thermal fluid. A temperature sensor placed near to the outlet pipe at the top of the collector relays temperature information to a separate controller unit. In turn, the controller activates a pump when it knows that there is a significant temperature difference (usually between 6°C and 10°C) between the top of the collector and the water close to the heat exchanger within the dual coil hot-water tank.

Evacuated tube collectors

As the name suggests, evacuated tubes are collectors that have had the air evacuated out of them to prevent heat-loss through convection. This design significantly reduces heat-loss to the surrounding air and has the advantage that hot water can be produced even when the outside temperature is sub-zero. There are several designs of evacuated tubes. One well-established evacuated tube manufacturer is Thermomax, whose tubes contain a sealed copper pipe (known as the heat pipe)

which is filled with a non-toxic antifreeze just like in the pipes of a flat-plate collector. Each heat pipe has a copper fin attached to it which stretches across the diameter of the tube (known as the absorber plate).

Thermomax absorber plates also have a selective coating with the same properties as flat-plate collectors, in order to eliminate most heat loss through radiation. Protruding from the top of each tube is a metal tip which is attached to the heat pipe (the latter being the heat exchanger in these systems). As the sun shines onto the selective coating of each absorber plate, the antifreeze within each heat pipe is heated causing it to evaporate and rise to the top. Here, the copper tip connects with a header pipe through which more antifreeze flows.

Figures 4.8a, 4.8b and 4.8c c (a) Display at the Centre for Alternative Technology, UK, showing on the indirect (left) and direct (right) evacuated-tube systems; (b) and (c) Close-ups of evacuated tube collectors

Source: (a) & (b) Frank Jackson (c) BSW-Solar/Langrock, http://en.solarwirtschaft.de

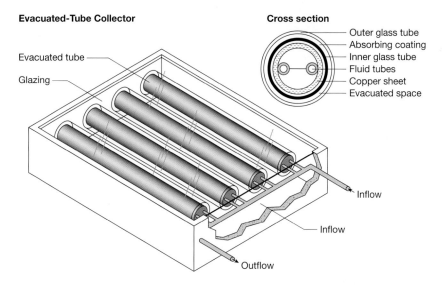

Figure 4.9 Evacuated tubes contain vacuums to retain more heat. Fins, or a cylindrical surface that maximises the area of collector directly facing the sun, are attached to, or enclose, the fluid-containing tubes. This increases the period over which heat can be collected, both during a day and throughout the year. The surface is given a coating, which absorbs more heat radiation frequencies. Evacuated tubes are more efficient than flat-plate collectors mainly because they do not re-emit as much heat thanks to the vacuum within the tubes.

Source: US Department of Energy, from the *Earthscan Expert Series* book *Solar Technology* by David Thorpe.

Figure 4.10a, 4.10b, 4.10c and 4.10d
Mirrors can be used inside or outside a tube collector

Source: (a) Chris Laughton (b) Solfex (c) Ritter Solar/ESTIF (d) Schott-Rohoglass/ESTIF. From the *Earthscan Expert Series* book *Solar Domestic Water Heating* by Chris Laughton.

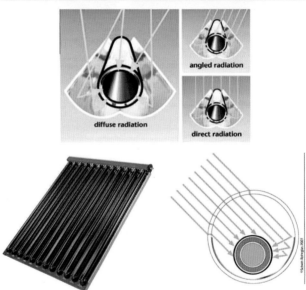

As with flat-plate collector systems (see below), a temperature sensor is attached to the header pipe to relay information that enables the controller to activate the pump when the correct temperature difference between the header pipe and the heat exchanger within the hot water tank is reached. In evacuated tubes the heat is transferred from the copper tip to the antifreeze within the header pipe, then the vapour within the tip liquefies as it cools and returns to the bottom of the pipe to repeat the cycle. There is a constant process of transferring the sun's heat during the day into the hot water tank.

Integral collector storage systems (ICSs)

ICSs are generally fitted into a glazed and insulated box. All ICSs function as a hot water pre-heat and pre-heat store for the solar-heated water and they are usually plumbed directly into the property's conventional hot water cylinder. Integral storage systems can be passive, but this only works well in regions unaffected by freezing winters. They function particularly effectively if there's heavy demand for hot water during daytime and early evenings.

Figure 4.11 A Sunflow System with integral storage. Within the casing a reflective mirror concentrates heat onto a stainless steel collector/cylinder containing 25 gallons of water. The heated water from the collector is drawn from the top of the first collector to the bottom of the second, where it is raised to an even higher temperature.

Source: Hydro Quest/American Viking Water heaters, http://servamaticsolarparts.com

Dual Coil – Solar Compatible – Hot Water Tanks

Most new solar water-heating installations will include replacing a traditional (single coil) hot water cylinder with a specially designed and super-insulated solar cylinder, which is integral to optimising system efficiency. This should be suitably sized for the area of panelling so as to maximise its ability to store the energy harvested by the collectors, without being so big that the water rarely gets beyond lukewarm.

A good rule of thumb is 55 l/m^2 of flat-plate panelling (the US DOE suggest 1.5 gallons per square foot of collector). A hot-water tank that is too large will have difficulties reaching the required hot-water temperatures. If there's a particularly long pipe running between the solar absorbers and the hot water tank, this can add hot-water storage to the system, so long as the pipes are well lagged. In this situation, it may be possible to get away with a slightly smaller tank. It's also possible to have two cylinders: the original ordinary one and an additional solar tank. The solar tank will receive the solar-heated water and pass it on to the main household tank as required (e.g. when demanded by the system users or if the solar tank gets close to boiling). 'Approved' systems are designed to ensure that the water coming out of the tap never gets dangerously hot.

The tank has two heat-exchanger coils:

- one at the bottom which is connected to the solar circuit; and
- another at the top which is connected to a conventional heat source (e.g. a biomass or fossil fuel boiler or back boiler).

Even on days when the sun is not shining, water is preheated by the solar circuit coil. With plenty of solar energy available (typically during late spring and summer), the solar hot-water system will be able to heat the water in the cylinder up to the desired temperature all by itself. On days when there is less than enough solar energy available, the conventional heat source can be used to top-up the temperature to the desired levels. Ideally, the solar water cylinder needs to be tall

Figure 4.12a
Cutaway of a twin-coil solar hot-water tank showing a simple layout for stratification, without sensors

Source: David Thorpe, from the *Earthscan Expert Series* book *Solar Technology*

Figure 4.12b
An Okofen twin-coil solar hot-water tank, with a section removed to show the tank's wall-thickness and the two coils within

Source: Okofen, www.pelletsheizung.at

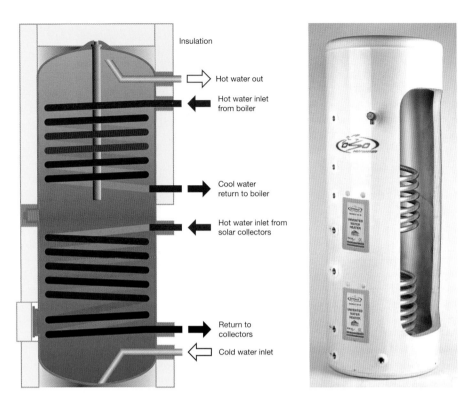

and thin so as to maintain what is known as stratification: the natural layering of less-dense warmer water over cold water, resulting in the temperature towards the top of the cylinder being several degrees higher than that near the bottom. Effective heat exchange is maximised by ensuring that the solar circuit can provide heat to the lower temperature water in the tank.

Heat exchangers for solar circuits need to have much larger surface areas than conventional heat-exchanger coils. Typically, copper-finned heat exchangers (rather than a smooth bore steel type) are used for solar circuits since they have coils with higher U-values (see Chapter 2), meaning that less surface area is required. When water in a hot-water cylinder remains between 20°C to 46°C for prolonged periods there can be a risk of bacteria growth (such as legionella). National standards and codes will cover this issue and commercially installed systems will comply with these standards. To prevent the risk of legionalla proliferating, some codes specify that the timer of the conventional heat-source is programmed to heat the water to 60°C (140°F) every 24 hours to kill off any bacteria. This should ideally be done towards the end of the day when the system is still hot but when the sun can no longer provide any radiation to the panels, in order to minimise the use of conventional fuels.

Types of Solar Thermal Systems

There are two main types of solar water-heating systems: active, which have circulating pumps and controls, and passive, which don't. Generally speaking, thermo-siphon systems (see below) are passive and most others are not. As well as active and passive systems, there are also direct and indirect systems.

Direct Systems

A direct system does not need a heat exchange unit since the water consumed by the household or business is directly heated within the collector. This type of system is not considered suitable for climates where sub-zero temperatures are likely to occur as this will damage the collector and, since the system is direct, you can't use anti-freezing fluid. An additional problem is created by mains water being fed into a collector loop, causing limescale buildup and consequently shortening the life of the system. For these reasons, direct systems are less common.

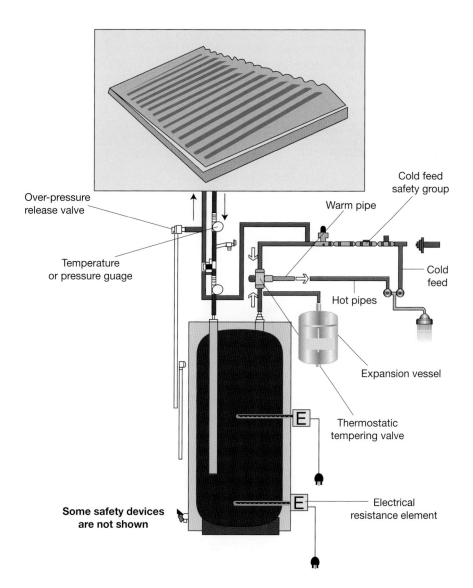

Figure 4.13
Schematic of a typical integrated collector storage system with electrical backup

Source: Chris Laughton, from the *Earthscan Expert Series* book *Solar Domestic Water Heating*

Indirect Systems

Most modern solar thermal systems heat the water indirectly through a heat-exchange coil within the hot-water tank. There are three main types of indirect system:

- fully filled;
- drain-back; and
- thermo-syphon.

Fully filled system

Fully filled systems have a constant flow of liquid throughout the loop between the collectors and the coil. To prevent frost damage to collectors, the thermal liquid used is invariably an antifreeze solution. This system is ideal for evacuated tube arrays, which always use antifreeze in the mix. Fully-filled systems can also be used with flat-plate collectors provided that they are also filled with antifreeze solution.

Fully filled systems require a number of components that are also common to many other solar thermal heating applications.

Other system components

Arguably the heart of a solar water-heating control system, *differential thermostats* measure the difference between collector and storage cylinder water temperatures. When the collectors are 6°C to 11°C warmer than the cylinder, the thermostat switches a pump (or fan for air systems) on to circulate the heated water (or air) to where it's needed. A *temperature sensor* is attached to the header pipe to relay information to the controller, which activates the pump when there

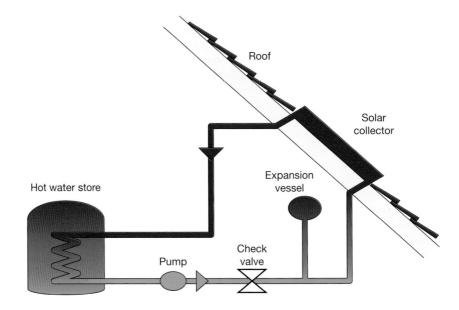

Figure 4.14 A fully filled, indirect sealed solar thermal system layout. It includes an expansion vessel to contain any expansion of the fluid when it gets hot.

Source: David Thorpe, from the *Earthscan Expert Series* book *Solar Technology*

is a significant temperature difference between the header pipe and the water close to the heat exchanger within the dual coil hot water cylinder.

This is usually a *one-way valve* to prevent the return flow between the pump and collector, thus preventing the cylinder from being cooled down by cold solar liquid from the collector when the pump is not running (i.e. any time when the sun is not providing heat to the collector, such as overnight). *Expansion vessels* – or membrane expansion vessels – are basically a safety mechanism that absorbs temperature-related volume changes in the antifreeze solution which would otherwise build up too much pressure within a sealed loop (indirect) system. The expansion vessel contains pressurised nitrogen, an inlet for the antifreeze solution and a membrane that separates the antifreeze solution from the nitrogen. As the pressure increases due to temperature rises in the antifreeze, any excess volume of liquid enters the vessel. As it does so, the increased pressure in the system pushes on the membrane with the nitrogen effectively cushioning the expansion. An additional safety valve – an *automatic pressure release valve* – is installed in the system to discharge any excess pressure that the expansion vessel is unable to handle. The excess antifreeze solution needs to be expelled somewhere safe where it can't harm anyone or cause any damage.

Thermostatic mixing valves are particularly important for evacuated tube systems, and are built into solar water-heating systems to protect against dangerously over-heated water coming into contact with people in the building. On a hot summer day it is possible for the temperature within the cylinder to reach 90°C, so it's essential to include thermostatic mixing valves at the hot-water outlet of the hot-water cylinder. If the water is too hot (say above 60°C) the valves mix cold water with it and protect users from scalding.

Drain-back system

In most systems, when there is no heat available from the sun or when the water in the cylinder reaches the required temperature, the pump stops. With drain-back systems, after the pump has stopped the water drains back into the drain-back vessel (a non-pressurised tank) using gravity. This can protect the collector from frost damage. A drain-back system also protects the system in the case of a black-out since this means that electricity isn't required to drain the panels. The pump refills the collector when the sun's heat is detected.

Thermo-syphon system

Instead of a pump transferring warmer thermal fluid from the collector into the heat exchanger, a thermo-syphon system relies on the natural movement through convection of fluid due to differences in fluid density at different temperatures. As the thermal fluid is heated in the collector it rises naturally out of the top into the tank, which must be located above the collector. Via the tank's heat exchanger, the heated fluid is transferred and the thermal fluid becomes dense again and flows downward through a pipe into the bottom of the collector to be reheated again. However, building layouts often make it difficult, if not impossible, to have the tank above the collectors. In this situation, a thermo-syphon system would not be possible and a pumped system is the only option.

Figures 4.15a and 4.15b Diagram and photo of a thermosyphon system with integrated storage tank

Source: (a) US Department of Energy, from *Earthscan Expert Series* book *Solar Technology* by David Thorpe, (b) Copyright @ 2005 David Monniaux http://commons.wikimedia.org

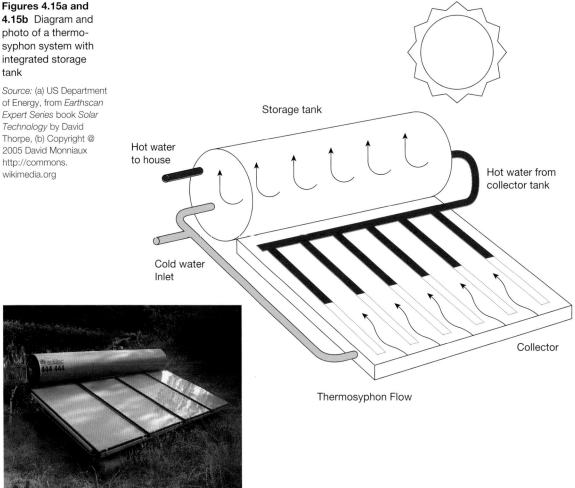

These types of systems don't require a pump and, because of this, are sometimes called passive solar systems. They are very robust and generally require little maintenance providing that they have been designed well in the first place. Compared to pumped systems, however, they are not quite as efficient.

Selecting Appropriate System Type

Many solar water-heating system suppliers tend to have their own preferred systems and makes, so to get a wider range of opinions it makes sense to discuss your hot water needs and existing hot-water heating system with more than one supplier. The relative advantages of different system types have been discussed above and can be weighed alongside the advantages of a local supplier–installer who can also offer aftersales servicing at some future date.

Once the system type has been selected, the system size has to be determined. Again, your supplier–installer will be able to advise on this, but a general rule of thumb for sizing a solar hot-water system is that there should be 1 m² of flat-plate panelling (less for evacuated tubes) per person in the household. Something to be avoided in a solar-thermal system is stagnation of the thermal fluid. This liquid is a biodegradable glycol based alcohol, and if it overheats within the collector it may cause it to stagnate, resulting in the degradation of the thermal fluid which causes residues to be left behind thus shortening the life of the collector.

If the water in the tank is already at its maximum temperature, then the pump will stop pumping the thermal fluid from the collector into the heat exchange coil within the tank. This can be overcome by sizing the tank according to the area of panelling. A volume of 55 l for every m² of flat-plate panelling installed in a temperate region and a maximum tank temperature of 60°C would accommodate the storage of the energy from a full day's sunshine whilst avoiding stagnation of the thermal fluid.

In hotter regions the volume may need to be increased. If available space for the hot water tank is an issue, this can be accommodated by setting the maximum temperature to between 80°C and 90°C, thus allowing it to store more of the sun's energy with less volume. One drawback to this solution is that the heat exchanger will be more prone to limescale buildup. If the tank has reached its full heat-capacity and the collector is still receiving solar energy, then a heat-dump module is recovered to discharge any excess hot water.

The Economics of Solar Water Heating

Solar water-heating systems do cost more for the equipment and installation than conventional water-heating systems. However, they can end up saving money in the medium- to longer-term because of their exceptionally low running-costs based on essentially 'free' heat from the sun. The payback and exactly how much money will be saved depends on a number of factors:

- The amount of hot water that the household consumes;
- The system's performance (which is largely determined by system design, the solar resource available and positioning of the collectors);
- Available incentives in your country of residence; and
- The cost of the conventional fuels being replaced (gas, electricity, oil or wood), the cost of which is generally rising.

A typical system installed in sunny regions of the US or Australia could cover up to 80 per cent of an average household's hot-water needs over the course of a year. In the UK, this is more likely to be between 40 and 60 per cent. While the sun's energy costs nothing, fossil fuel prices continue to rise as and reserves are dwindling, so installing a solar thermal system is one way to protect against future fuel shortages and price hikes.

Solar-assisted Space Heating

'Passive' solar space-heating is covered in a separate section below. Here we look at 'active' solar-assisted space heating. By 'active' we mean that it involves the use of powered devices, like circulation pumps or fans. There are two types of active solar space-heating: one uses liquid and the other uses air as the principal heat-transfer substance. In both cases there's a solar collector that feeds solar heat into a building. Taken together, space and water heating can account for up between 12,000 and 20,000 kWh/yr even in domestic situations. In temperate regions, space heating is likely to demand at least four times as much heat as the hot-water element. This ratio will be different for a business that needs hot water for processes (e.g. in food preparation).

In the case of liquid-based systems, the heat is often transferred directly into the building (as with air) or it can arrive via a heat-storage device, such as a solar cylinder or buffer tank, from where it can be diverted as needed into the property's wet heat-distribution system. Liquid systems are plumbed in. Larger air-systems need ducts, while smaller – one room – systems can be mounted to external walls and may not need ducting. Whatever the scenario, it only makes practical sense where the building is very well insulated and has low heating requirements as well as being exposed to sufficient winter sunshine.

Space-heating systems are only very rarely designed to meet 100 per cent of a building's space-heating needs. To do this would not be very cost effective, so they are generally matched to a backup heating source (any kind would do: conventional, biomass-based or even heat pump) – hence the term 'solar-assisted'. Consequently, most systems are designed to meet the property's average winter heat-load rather than its peak load. This is significantly more cost-effective since it avoids having to build a much larger solar collector, which would be oversized for the majority of the year. Nevertheless, some properties do go a long way to providing hybrid solar water- and space-heating. The collector area needed for a single-family solar hybrid household in Austria is calculated to be between 12 and 20 m² (i.e. 129 and 215 ft) which would provide around 40 per cent of the total heat demand if the house were well insulated. There are even district heating systems that have a solar-assisted heating input, usually based on a single large building, but also for systems serving several buildings.

Solar heat is also used for processes such as crop drying. This can be a passive system, some as simple as a glass or plastic roof, or an active one, for instance

Figure 4.16 Solar collectors for space heating on a Passivhaus-standard office building in Neudorfer in Rutzemos, Austria

Source: © IEA-SHC, from the *Earthscan Expert Series* book *Solar Technology* by David Thorpe

where warmed air is circulated by fans through ducts. Solar heat is sometimes used to preheat water or air for industrial processes. In these situations the water or air temperature will usually be topped up to achieve the higher heats required for the process. The financial savings can be significant due to the solar pre-heat and minimal use of fossil fuel top-up.

Solar Liquid Heating

Solar collectors that utilise a liquid medium work better than air-based systems when connected to a property's central-heating network. Usually flat plate or evacuated tubes (the same as are used in solar water-heating systems) are used to optimise solar collection. Heated liquid flows from the collectors either into a storage tank for later use or to a heat exchanger for immediate use. Like standard solar water-heating installations, the system also requires a circulation pump, expansion vessel, valves, heat exchanger and storage tank. Unless the building manager has all the necessary skills and experience, liquid solar heating-systems are best installed professionally. This is, in any case, often a requirement for any grants or renewable heat incentives.

Space heating with liquid-based solar collector systems is a significant investment, so it's vital to size the system and calculate the necessary flow-rates as accurately as possible. Again, this is an expert task and is best left in the hands of professional suppliers–installers. In terms of distributing the liquid heat, the most efficient systems are: underfloor radiance and heated skirting-boards or conventional wall-mounted radiators (see Chapter 2 for more details on these).

Figure 4.17 An evacuated-tube system that provides domestic hot water as well as radiant underfloor heating. Note the large number of solar collectors needed for the space heating – several times that which would be needed for domestic hot water alone.

Source: Thermomax Industries www.solarthermal.com

The most effective is underfloor, since this works efficiently at relatively low water-temperatures.

Solar-assisted heating systems are particularly suited to regions with heating seasons that are long and cold, as these offer greater opportunities for recovering the initial investment. By matching the average winter heat-load to the rated output of an active solar collector system, it's simple enough to work out how many collectors will be needed. A solar-assisted space-heating system that has to meet the very coldest of winter days would require a massive and very expensive set of solar collectors which would be vastly oversized during the rest of the year. It's more economical to use some kind of backup boiler to meet the peak heat demand. This could be a biomass stove, gas boiler or any conventional type that can accomodate wet heat-distribution, depending on circumstances and preferences. System efficiency is optimised by using a liquid-based active solar space-heating system for a building's hot water needs too so that it gets more summer use.

5
Heat Pumps

Basic Principles

Heat pumps are not a new technology, in fact most of us already have a heat pump in our home in the form of our standard refrigerator. Heat pumps for domestic heating work like a refrigerator turned inside out: instead of removing heat from a space and dumping it into the surroundings (which is why the outside of your refrigerator is always warm) heat pumps absorb heat from the surroundings (either ambient air, ground or water) and, using a small amount of electrical energy, change it into useful heat energy which is delivered into your home. They're popular for use in very energy-efficient buildings and there are over 500,000 installations in the US and 35,000 in Canada.

We don't usually think of our surroundings as containing heat, but energy from the sun's rays is constantly being absorbed and stored in the air, ground and water. Even on the coldest of winter days there is abundant heat energy all around us: it is just so dispersed that we are usually unable to make use of it. One illustration of this is the phenomenon of urban 'heat islands' where the accumulation of low-grade heat given off by buildings, human processes and the sun makes the temperature a couple of degrees higher in cities than in the surrounding countryside. The ground below us, too, absorbs roughly half of the sun's energy and so possesses low-grade heat in most of the world's habitable environments. In the UK the average ground temperature at 10 m below ground-level varies from 10°C to 14°C.

The purpose of heat pumps is to absorb dispersed low-grade heat, upgrade it to a higher quality and deliver it to the space that needs heating (e.g. your home). Low-grade heat is considered by many as both renewable and essentially free. All that is required is the input of electricity to run the pump and the pump itself. As such, the electricity required does raise some issues: for a heat pump to be a fully renewable and an effectively carbon-zero heating source, clean electricity from a renewable source has to be used.

Heat-pump efficiencies are often described in terms of the Coefficient of Performance (COP), which describes how many units of energy a device delivers for every unit that it uses. A heat pump that delivers 300 per cent efficiency (i.e. that delivers three units of heat for every one unit of energy – commonly electrical energy – put in) would have a COP of three, whereas a standard electric heater has a COP of 1.

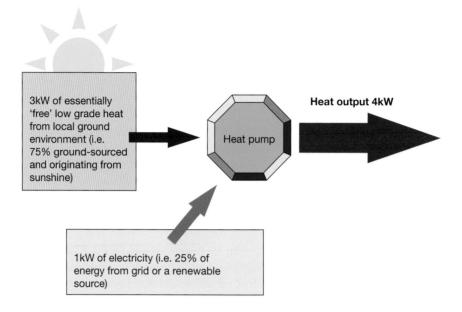

Figure 5.1 Coefficient of Performance (COP) for heat pumps is usually between three and four; in this example we show it as four, where 75% of the energy comes from sun-heated ground source and 25% is electrical input, either from grid or renewable sources

Source: Dilwyn Jenkins

The basic principle of heat pumps is heat transfer. Heat naturally flows from hotter temperatures to colder ones (which is why on a warm day your room heats up as the heat from outside flows in, while on a cold day your room cools down as heat from inside flows to the cooler outdoors). Heat pumps use electricity to reverse this process, so rather than creating heat – for example by burning fossil fuels – heat pumps simply move heat from one place to another using a heat conducting liquid called a refrigerant.

How Heat Pumps Work

Like all liquids, refrigerants boil and condense at different temperatures and pressures. They absorb heat as they boil and release heat when condensing. Heat pumps are designed to exploit these physical properties. Different refrigerants are used to meet the specifications of different systems, but in general refrigerants must have a low boiling point and they must be chemically stable and non-corrosive. Prior to 1989, CFCs (chlorofluorocarbons) and HCFCs (hydrochlo-roflurocarbons) were commonly used. However, in accordance with the Montreal Protocol, these have now been replaced by more ozone friendly substances such as HFCs (hydroflurocarbons).

A heat pump is essentially a sealed system of pipes containing refrigerant with two heat-transfer points: one where the low-grade heat is absorbed (the **evaporator**) and one where the higher-grade heat is delivered (the **condenser**). Heat pumps are made up of two sides: a high-pressure and a low-pressure side. The liquid in the low-pressure side enters the **evaporator** where it absorbs heat from the low-grade source using a **heat exchanger.**

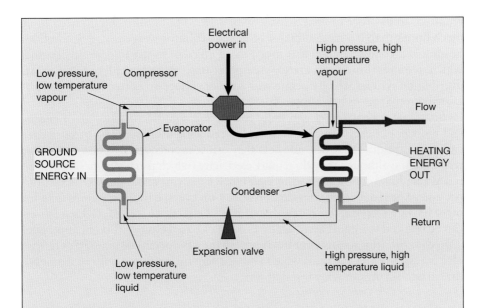

Figure 5.2 The heat pump cycle uses a combination of ground-source and electrical energy to transform a liquid from low pressure and low temperatures into a high pressure and high temperature vapour

Source: Dilwyn Jenkins

A heat exchanger is simply a device with a large heat-conducting surface area which allows for the quick exchange of heat (a good example is a household radiator which transfers heat from hot water into the surrounding air via metal plates). Sometimes the heat exchanger picks up heat directly from the source. However, it is usually connected to an external closed-loop system of pipes containing a mixture of water and antifreeze that absorb the heat from the source and transfer it to the pump (options for heat absorption are discussed further on in this chapter).

The refrigerant has a very low boiling point and so contact with this low-grade heat transforms it from a liquid into a gas. This change of state or 'phase change' (evaporation) is endothermic, meaning that a great deal of heat from the surroundings is absorbed in the process. The gas then enters the **compressor**, which increases the pressure of the gas resulting in an increase in temperature. The high-temperature gas then enters the **condenser**, where it cools and changes back into a liquid again. This change of state (condensation) is exothermic, meaning that a substantial amount of heat is released in the process. Another **heat exchanger** then delivers this heat into your living room or hot taps by transferring the heat to a boiler or fan system. The liquid then flows into the **expansion valve** where its temperature and pressure are lowered and the cycle begins again.

In a temperate climate like much of the US and UK, a heat pump installed in a well-insulated home can be expected to deliver between three to five times more energy than it requires to run, making it 300 to 500 per cent efficient (this works out at around three to five times more heat per unit of electricity compared to a standard electric heater). If a home or business makes significant use of electric backup heaters to 'boost' the output in cold weather the COP will be significantly reduced. In order to achieve higher temperatures inputs of greater energy are required, so in order to heat a space with a small radiator it would need to be much hotter than if a large radiator or heat distributor were utilised. Underfloor heating is a good example. It is much more efficient to use underfloor heating, which requires lower circulation temperatures, than typical radiator heat-distribution systems which operate with smaller surface areas (see section on Heat Distribution Systems in Chapter 2).

Installing a heat pump can help you save a significant amount of energy and money, but they can also be useful in reducing a household's or business's carbon footprint. The low-grade heat they utilise is considered to be a renewable resource, and although they do require electricity (which usually comes from non-renewable fossil-fuel sources) the high efficiencies can make the CO_2 emissions per unit of heat delivered considerably lower on average for heat pumps than for other heating systems. If a renewable source of electricity is used, CO_2 emissions can be reduced to almost zero. Many fossil-fuel-based electricity companies offer a 'green tariff' (this means your bill is paying for a proportional amount of renewable energy to enter the grid) which you can sign up for if you wish to maximise the carbon savings of your installation.

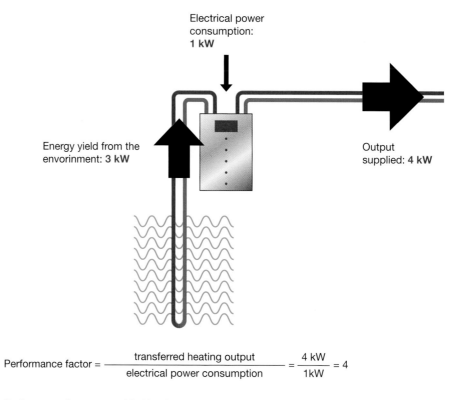

Figure 5.3 Coefficient of Performance factors – in this example 3 kW of heat is extracted from the Earth and 1 kW of electricity is put in from the mains to achieve an output of 4 kW of heat, offering a coefficient of performance of four

Source: Viessmann, www.viessmann.co.uk

$$\text{Performance factor} = \frac{\text{transferred heating output}}{\text{electrical power consumption}} = \frac{4\text{ kW}}{1\text{kW}} = 4$$

Performance factor = provided by the manufacturer, lab. value to EN 255
Annual per. factor = ratio of energy yield -v- energy spent over a 12 month period

Heat pump system type	Seasonal efficiency
Direct GSHP system with under floor heat distribution	300% to 400%
Indirect GSHP system with under floor heat distribution	350% to 500%
Air source heat pump	250%

Figure 5.4 Seasonal efficiencies of heat pump types: ground source heat pumps (GSHP) are more efficient than air source heat pumps (ASHP). Note that it is very rare to achieve a seasonal efficiency of 500% (which is the same as COP 5).

Source: Energy Saving Trust, www.est.org.uk

It is important to note that the environmental performance of a heat pump installation is highly dependant on how well-insulated the building is and the type of heat-distribution system in the building. A recent field trial by the Energy Saving Trust (2010) found that many UK heat-pump systems are underperforming, with an average COP of just 2.4 for ground-source heat pumps (different types of heat pumps are discussed in detail later on in this chapter). Figure 5.4 shows both the relative efficiencies of common heat pump types (including air source) as well as comparing the seasonal efficiencies, since the COP changes along with climatic conditions during the year.

The seasonal efficiency of a heat-pump system is the ratio of the energy delivered from the heat pump to the total energy supplied to it during a year of operation. This ratio is derived from the Coefficient of Performance, but remember that the calculation also includes the energy required for fluid circulation around the heat exchanger.

Choosing the right type of heat pump

A number of variations on the heat pump are available, which make use of different combinations of heat source, absorption medium and heat-distribution medium. Your choice will depend on the heat sources available to you, the space available to you, the heating and cooling demands of your home and its size.

Common types of heat pump:

- Ground to Water;
- Ground to Air;
- Air to Water;
- Air to Air;
- Water to Water; and
- Water to Air.

Open-loop or closed-loop (sometimes called direct or indirect)

Heat pump systems can either use open-loop or closed-loop systems. A closed-loop system of pipes is designed to absorb heat from the source and then transfer it to the pump. With open-loop (or direct) exchangers the pump refrigerant absorbs the heat directly from the source. However, some open-loop systems have an intermediate heat-exchanger (e.g. this might be chosen if source water is corrosive, since it is better to damage an intermediate heat exchanger than spoil the heat exchanger's evaporator). Most engineers use the open-loop/closed-loop distinction, since an indirect open-loop is possible.

Electrical connection

Heat pumps are operated by an electric motor which can cause disturbances to the electricity-distribution network because of high starting-currents. This problem is worse for single-phase connections and can lead to flickering lights and voltage surges (which may affect electronic equipment). The heat-pump installer will be able to advise on these issues, and may contact your distribution-network operator to find out exactly what the maximum load that can be powered from the grid at your project location is. For large systems you may need to upgrade your electricity supply from single- to three-phase. Other possible technical fixes include soft start controls, multiple heat-pumps (e.g. for different floor levels) or the use of systems with low-torque compressor starting. If you do upgrade or change your supply, there are obvious environmental benefits to buying it from a supplier of renewable or green electricity.

Reverse-cycle heat pumps

These are designed to allow you to switch between heating or cooling depending on the season. The principle is exactly the same for both heating and cooling: it is simply a valve allowing the direction of flow to change (whether heat is being taken into the building or out of the building). Both ground-source and air-source heat pumps can be reverse cycle.

Simultaneous heating and cooling

Some systems can be designed to offer heating and cooling simultaneously. Such systems move heat from areas where it is unwanted to areas where it is needed, although there is rarely a good match between cooling and heating loads. Where it does match well, simultaneous systems can be very efficient, as heat that would otherwise be dumped outside by a conventional air-conditioner can be utilised to meet heat demand elsewhere. For optimum efficiency this type of system requires

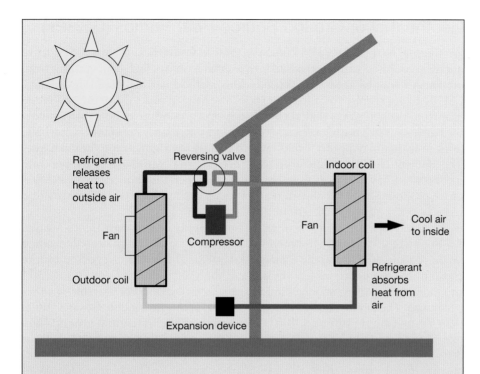

Figure 5.5 An air source heat pump providing summer cooling. Heat from the air inside the building is absorbed by the refrigerant and released to the outside.

Source: Dilwyn Jenkins

the heating and cooling demands of the building to remain in equilibrium for the majority of the time, otherwise in times of excess cooling-demand heat will have to be dumped, and in times of excess heating-demand an additional boiler will be required.

Reverse-cycle chillers

One advanced innovation in air-source heat pumps is the reverse-cycle chiller (RCC), which allows building managers to choose from a range of heating and cooling distribution-systems. RCCs are economical for all-electric homes and in some circumstances will be the least-expensive heating option. They consist of a standard single-speed air-source heat pump which is sized to the heating load (not, as is common, the smaller summer cooling-load). The heat pump is connected to a large and well-insulated water tank, which is either heated or cooled by the heat pump, depending on the season. Fan coils and ducts distribute the stored hot or cool water to the building. For heating, they are most efficient in combination with a radiant underfloor system.

Geothermal or Ground Source Heat Pumps (GSHPs)

Geothermal heat-pumps, also known as ground source heat pumps (GSHP), are not to be confused with geothermal power, which utilises high temperatures from the Earth's core to generate electricity (this method is widely used in Iceland). GSHPs do not require so-called 'hot rocks' but instead make use of the low-grade heat from the sun's rays that is absorbed by the ground. Around a metre below the Earth's surface the temperature tends to stay at a fairly constant 8°C to 12°C (even in winter in temperate zones) which is more than sufficient to heat a house. The ambient climatic temperature, and even diurnal variations in temperature (i.e. between day and night), will affect that temperature noticeably in the first metre of ground. GSHPs absorb the heat via a ground loop which consists of a series of pipes buried in the ground.

Ground loops can either be installed horizontally in a shallow trench around 1.5 m down, or vertically in boreholes between 15 and 100 m deep depending on the space available and the geology of the area. The deeper one goes, the more constant the temperatures are throughout the year. Vertical loops are used in situations where there may be limited space, rocky ground or extreme weather conditions. The depth required depends on local substrate conditions, rarely more than 70 to 100 m (100 to 300 ft). The length of the horizontal ground-loop that is required will depend on the size of the space being heated and the heat demands. A poorly insulated house would require a much greater length to maintain a sufficient temperature, as a significant proportion of the heat produced will

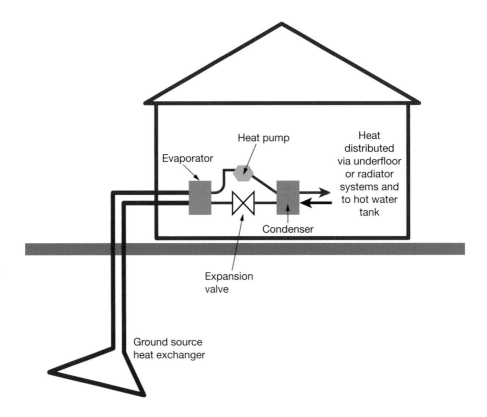

Figure 5.6 Diagram showing the main components of a heat pump system. In this ground source system heat is taken from beneath the surface of the ground and transferred to a household hot-water system. The heat pump itself runs off electricity.

Source: Dilwyn Jenkins

HEAT PUMPS 99

Figure 5.7a and 5.7b Diagram showing two different configurations for a ground source heat pump. The first (a) shows a horizontal system where the pipes are buried in a shallow trench. The second (b) shows a system where the pipes are installed vertically.

Source: Bundesverband Wärmepumpe e.V., Germany, www.waermepumpe.de

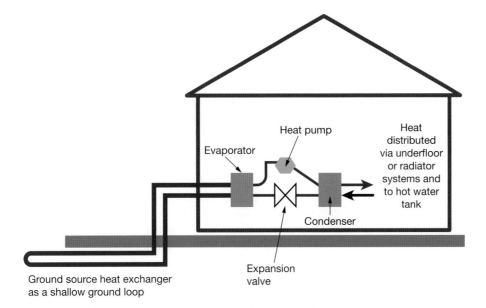

Figure 5.8a GSHP heat exchanger with shallow sealed-loop system. The pipes are laid horizontally in shallow trenches which are then back-filled.

Source: Dilwyn Jenkins

Figure 5.8b GSHP heat exchanger with 'slinky' shallow coil. The trenches are filled with coiled pipe which allows more pipe to be installed in a smaller area.

Source: John Cantor, Heat Pumps, www.heatpumps.co.uk

Figure 5.8c Slinky coils laid out on the surface before they are installed in trenches.

Source: Mark Johnson http://en.wikipedia.org

be lost through draughts. This inevitably increases the costs of the installation substantially.

A typical horizontal installation will require between 100 and 600 m of piping. This can be laid either as straight horizontal pipes or as a coiled pipe (known as a 'slinky') which enables the trench size to be reduced without significantly decreasing heat production. That said, there is only so much heat in a specific area of ground. A typical vertical installation will require 80 to 100 m of piping, although this need not be one borehole of 100 m but can be installed as several shallower boreholes. One borehole should be sufficient for a low-energy house, while more are likely to be required for less efficient or larger properties. Businesses can also use larger borehole systems towards building and process heat-needs. Boreholes can also enable GSHPs to be installed in relatively small spaces. However, they do tend to be more expensive to install than many other renewable energy systems. It is also important to make sure that the geology allows for the digging of boreholes, and that ground-water sources will not be adversely effected. Carrying out a geological survey may be necessary (though sometimes asking your local digger-driver for local geological information can be sufficient) before commencing digging, and may be advisable in order to avoid unforeseen costs.

The most common system is the indirect-circulation system consisting of a closed-loop series of polythene pipes filled with a mixture of antifreeze and water which absorb the heat and transfer it to the refrigerant via a heat exchanger. Alternatively, heat can be transferred using a direct expansion system in which the refrigerant circulates through a series of copper pipes, absorbing the heat from the ground directly. These systems are very rare, and while they are sometimes more energy-efficient they can also be more expensive and prone to refrigerant leaks. Generally, indirect and sealed loop-circulation systems are preferable. Once absorbed from the ground loop into the refrigerant the heat is then upgraded and transferred to your heating system via a heat exchanger.

GSHPs can either be classed as ground-to-water or ground-to-air systems. In ground-to-water systems the compressed and upgraded heat is used to heat water, which is then circulated around your home using a wet central-heating system. Ideally underfloor heating should be used as it provides the most efficient delivery of the kind of consistent lower temperature heat produced by GSHPs, although radiator systems can also work if designed correctly. As well as space heating, ground-to-water systems can also be used to meet household hot water needs. However, this can lower the efficiency of the system somewhat. Ground-to-air systems (common in the US but rare in the UK) transfer the upgraded heat directly into the air where it can be circulated using a fan system.

Benefits of GSHPs

Effective: GSHP can generate temperatures between 35°C and 55°C (ideal for use with underfloor heating).

Long life: a typical heat pump comes with a warranty (between two and ten years) and has an expected life of around 20 years while the ground loop can, if installed properly, last for up 40 years.

Low maintenance: they have been described as a 'fit and forget' technology because they require very little maintenance after installation. They are fully automatic and work quietly. Heat pumps will require an annual check which can be done yourself, and a check by a professional installer every two to five years.

Savings: GSHPs have some of lowest and most consistent running costs for heat pumps, which can help lower your energy bills considerably (although savings do depend on the system being replaced).

Carbon footprint reduction: GSHPs can lower a household's or a business's carbon footprint significantly, particularly if powered by renewables, which effectively zeros your heating footprint.

No visible external equipment: this is not only important from an aesthetic point of view but also limits damage from weather exposure, accident or vandalism.

Safe: the components are not explosive and there is no combustion involved, making them very low risk compared to heat sources such as gas and oil.

Clean: there are no flue or ventilation requirements and no local pollution produced.

Versatile: GSHP can be incorporated easily into many homes and can be designed to provide all of your household's heat and hot-water needs (a monovalent system), or to work alongside other energy sources in a bivalent system.

Usually no planning permission required: provided there are no special conditions such as listed building or conservation area status. Check with your local authority.

Can work well at scales appropriate for businesses: even business properties without much available land can use borehole GSHPs effectively for space heating and/or process water pre-heating.

Figure 5.9 Business tapping into underground low grade heat

Source: TCS Group / RHI Energies www.rhienergies.com

> ## Potential issues with GSHPs
>
> **Costs:** installation costs are generally higher than with other types of heat pump.
>
> **Space:** ground loops require a significant amount of space in order to absorb enough heat.
>
> **Geological barriers:** if bore holes are chosen, a survey to check the suitability of the local geology will be needed as this can affect the viability as well as the costs of a project.
>
> **Disruption:** the ground works will entail some disruption for both yourself and your neighbours, and the appearance of your garden may take a while to recover.
>
> **Groundwater protection:** the installation of a GHSP usually does not require permissions relating to excavation. However, in rare cases there can be risk to groundwater sources when boreholes are being used so it would be advisable to contact the relevant statutory bodies before beginning digging.

Suitability of GSHPs

To maximise efficiency and savings, GHSPs must only be installed in buildings that are suitable. The criteria in Table 5.1 should help you decide if a GHSP is right for your home or business premises:

Table 5.1 Criteria for selecting a ground source heat pump

Space	Sufficient ground space is needed to install the system loop and suitable ground for digging the trenches or bore hole as well as access for digging machinery.
Fuel being replaced	The savings you make will depend on the fuel being replaced. Savings made by switching from standard electric heating or coal will be much greater than if you are replacing a relatively new gas heating system.
Insulation	GSHPs produce heat at lower temperatures than conventional heating systems. Therefore it is essential that your home is well insulated if your heat pump is to function efficiently. Otherwise much of the heat generated will be wasted in leaks, making the system less effective, inefficient and more costly. Remember that installing a heat pump in a poorly insulated building can actually lead to higher energy bills.
Type of heating system	GSHPs work best when providing heat at lower temperatures (35°C is the most efficient) over a long period time, so are ideally suited to underfloor heating systems or large low-temperature radiators.
Starting up	When starting up, heat pumps require high electrical currents which can be an issue for some properties so it is always worth discussing any possible limitations with your installer. It is also possible to install heat pumps with a 'soft start' function that minimises this problem.
Timing	By combining the installation with the development of new building or renovation work, costs can be greatly reduced.

Horizontal or vertical GSHP placement	Horizontal systems are usually at least 1 m below the surface (to avoid frosting), with loops laid in sand and covered with about 15 mm of sand, but they need a lot more available ground surface area than vertical borehole types. Because of the need to drill deep into the earth, vertical systems tend to cost more (by approximately US$300 per kW) but generally offer a higher performance ratio. In terms of heat extraction rates per meter of the loop, vertical systems tend to be more effective (38 to 91 W/m as opposed to horizontals 38 to 59 W/m). Whether the system is horizontal or vertical, the buried part of the system should have a life expectancy of around 40 years and account for between 30 and 50% of the capital costs of installation. Most installer–manufacturers have software to help with system sizing.
Loop depths and spacing	It is better to bury the loop deeper to access more stable ground temperatures, which will increase both the collection efficiency and the installation costs. Horizontal loops tend to be installed at around 1 m below ground level (there may be regulations prohibiting access to unsupported trenches if they are more than a certain depth). Multiple pipes laid in a single trench should be at least 0.3 m apart and there should be a minimum distance of 3 m between trenches, if there is more than one. Vertical boreholes should usually separated by at least 3 m.
Soil and ground types	The overall heat extraction rate (calculated as W/m^2, or Watt per square metre) is also affected by the soil types and water content found on-site underground. Dry sandy soil is the worst (10 to 15 W/m^2). Wet loamy soil (25 to 30 W/m^2) and ground with groundwater (30 to 35 W/m^2) are the best. The thermal properties of the ground will need to be known in order to determine the length of heat exchanger (the loop or coil) needed to meet the specific heat load of a given property. Soil and rocks have significantly different values for thermal conductivity, with rocks being much higher. The moisture content of soil has a big effect, mainly because dry loose soil traps air and has a lower thermal conductivity than moist packed soil. EST (UK) suggest that low-conductivity soil may require as much as 50% more collector-loop than highly conductive soil.

Water Source Heat Pumps (WSHPs)

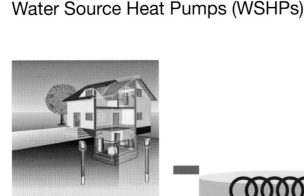

Figure 5.10a Diagram showing main components of a water source heat pump linked to a domestic underfloor heating system and hot-water tank

Source: Bundesverband Wärmepumpe e. V., Germany, www.waermepumpe.de

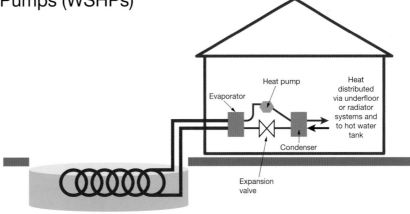

Surface water source heat exchanger coil

Figure 5.10b Diagram showing the main elements of a water source heat pump. Like all heat pump systems it includes a heat pump, an evaporator and a condenser – the difference here is simply the location and design of the coil loop.

Source: Dilwyn Jenkins

Figure 5.11 Slinky coils being immersed in a pond

Source: Mark Johnson, http://en.wikipedia.org

Water source heat pumps (WSHPs) follow the same principles as ground and air source pumps but make use of heat energy from the sun that is stored in bodies of water such as streams and lakes. They are also sometimes used directly (with open-loop systems) in wells, often purpose drilled, and frequently in pairs of wells where water is extracted from one to remove the heat, then returned to the other.

Usually WSHP use the *closed-loop* system where heat is absorbed from the water indirectly by pipes filled with water and antifreeze and transferred into the pump's refrigerant system via a heat exchanger. However sometimes an *open-loop* system can be used, in which water is taken up directly from a water body into the pump system, circulated and rejected back into the water body. Open-loop systems are however prone to corrosion and blockage, and permissions from the relevant statutory bodies are normally required, making closed loop systems a more appropriate choice in most circumstances.

Figure 5.12 Diagram showing a water source heat pump with a direct (open) heat exchanger. It is the water itself that runs into the pump system

Source: Dilwyn Jenkins

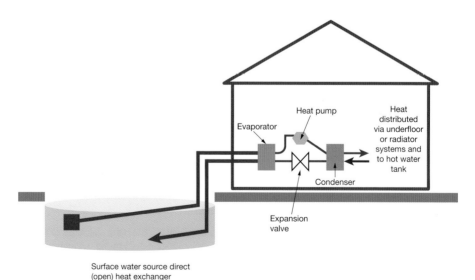

Benefits and suitability of WSHPs

WSHPs offer many of the same benefits as GSHPs, such as long lifespans (around 25 years), low maintenance and running costs, and carbon-footprint reduction, although the costs of installing a WSHP tend to be higher than for other types of heat pump. Installation also requires access to a fairly large body of water with a good depth and some flow in order to replenish the heat source. This obviously doesn't apply to all households or businesses. However, for those that do have this access WSHPs can provide an very clean, safe and efficient source of heating.

An abstraction licence or other permissions may be required for the installation of open-loop systems, since they involve pumping water from an existing source (river or pond) or via boreholes that are drilled to access water sources below ground for greater efficiency. Regulations may be less complex if the spent water is returned to the aquifer it originally came from. The basic issue is confidence that the system is not going to affect an existing user of groundwater in the vicinity or pose any other threat to the environment.

Air Source Heat Pumps (ASHPs)

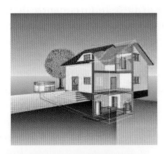

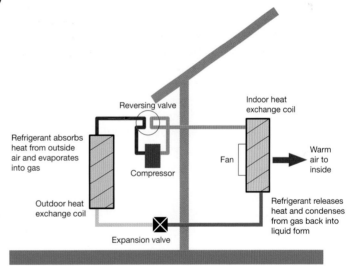

Figure 5.13a Schematic showing a house with an air source heat pump for space heating. The outdoor heat-exchange coil is housed separately from the building, connected to a wet heating system in the basement by underground pipes (air-to-water system)

Source: Bundesverband Wärmepumpe e. V., Germany, www.waermepumpe.de

Figure 5.13b Diagram showing the main components and possible layout of air source heat pump system, connected to a hot-air heating system (air-to-air system)

Source: Dilwyn Jenkins

Figure 5.13c Photograph of a domestic air source heat pump

Source: Energy Saving Trust, www.est.org.uk

What are they and how do they work?

Air Source Heat Pumps (ASHPs) absorb heat from the sun or waste heat from human processes that is stored in the air. There are two kinds of air source heat pumps: air-to-air and air-to-water. Air-to-air systems absorb heat from the air directly into the refrigerant via a heat exchanger unit, which can be either outside the building or in an internal space such as a loft. The heat is then upgraded by the pump and transferred via another heat exchanger into the air to be redistributed by fans, which can be installed as either a single-split or multi-split system. In a single-split system there is one external unit where heat is absorbed and one internal unit where heat is delivered, whereas in a multi-split system there are two or more delivery units that can serve different spaces in the building and can be independently controlled in many cases. Air-to-water systems transfer heat stored in the air into water, which is distributed via a wet central-heating system. As with GSHPs, underfloor heating is by far the most effective method of heat distribution for air-to-water heat pumps, although large low-temperature radiators can also be used. Air-to-water pumps are also often used for pool heating.

ASHPs can be installed on an external wall to capture heat from the air outside (and vice versa, or in reverse for cooling) or they can be installed internally as a heat recovery system. A heat-recovery heat pump absorbs heat from areas within your home which may have excessive unused heat such as loft spaces and kitchens, and transfers this heat to other rooms in the building where needed. This very efficient system can be used for both air-to-air and air-to-water distribution and is especially effective in small apartments.

ASHPs tend have the lowest installation cost of all heat pumps since the systems are simple and, unlike GSHPs, require no digging. Unfortunately, though, low winter air temperatures outside will cause lower winter operating efficiencies. Many people install a backup heating system to cover the coldest periods. Heat recovery ASHPs are available on the market that recover much of the heat that would otherwise escape the building in the exiting airflow, but these only work well in buildings that are airtight and have mechanical ventilation. As with all renewables, they can prove to be the most economical options and work extremely effectively in low-energy buildings.

Benefits of ASHPs

Long life: if well maintained ASHPs have a life expectancy of around 15 to 20 years, although it's fair to say that most contemporary split air-source pumps are considered quite antique at ten years or over.

Effective: ASHPs provide comfortable room temperatures, even at temperatures as low as 15°C.

Low maintenance: the only moving part is the fan and thus ASHPs can be described as a 'fit and forget' technology requiring very little maintenance after installation.

Easy to install: ASHP are much easier to install and entail much less disruption to your property than GSHPs, with air-to-air systems being the most simple of all to install.

Low capital costs: ASHPs are simpler to install and require fewer materials than GSHPs, making capital costs lower.

Savings: low running-costs can help to lower fuel-bills (although savings do depend on the system being replaced).

Carbon footprint reduction: an ASHP can lower your carbon footprint significantly. On average an ASHP could save five tons of CO_2 per year when replacing a standard electric heating system.

Safe: the components are not explosive and there is no combustion involved, making them very low risk compared to heat sources such as gas and oil.

Clean: there are no flue or ventilation requirements and no local pollution produced.

Versatile: because they do not require the ground-works that GSHPs do, ASHPs can be installed in a much wider range of properties. They can also be designed as part of a bivalent system to work alongside other technologies such as solar water heating.

Potential issues with ASHPs

Lower system-efficiency: overall efficiencies for ASHPs are inherently lower than GSHPs. Ground temperatures are generally higher than the mean air temperature in winter, so an ASHP would have to work harder to produce the same amount of heat. Similarly, ground temperatures are generally lower than the mean air temperature in summer, making cooling respectively easier. Ground temperature is also quite stable, whereas air temperature varies significantly throughout the day so the heat pump is often not working at its optimum temperature.

Defrosting: when operating at very low temperatures the evaporator coil will need defrosting. All units will have an automatic defrost mechanism, which temporarily diverts heat away from your distribution system and back into the evaporator. It can be advisable to have a hot-water buffer tank so that the building does not begin to cool while defrosting takes place.

Building code requirements: ASHPs are fairly new in terms of widespread public uptake and so may not yet be covered by the current statutory instruments. Installers will know what permissions are needed before installing an ASHP.

Noise: the main issue with regards to building code requirements is the fact that ASHPs do produce some noise. Noise levels vary between different units so it is a good idea to check the product specifications when choosing your unit, but a standard unit may produce noise levels in the region of 50 to 60 decibels (db) at a 1 m distance. To help you put this into perspective, a refrigerator normally operates at around 40 db while a washing machine operates at around 70 db, so a standard ASHP will register somewhere in between.

Air-to-air or air-to-water?

Air-to-air source heat pumps tend to be cheaper but less efficient and less versatile, in some ways, than air-to-water systems.

- *Costs*: air-to-air only requires a split reversible air-conditioning unit, making it simple and relatively cheap compared to air-to-water, which also requires a separate wet heat-distribution system.
- *Effective in large spaces*: fans help distribute heat effectively and more rapidly than wet heating-systems, which works well in larger spaces such as open-plan rooms, halls and offices. On the other hand, indoor air movements can make for discomfort.
- *Cooling option*: air-to-air units often provide both heating and cooling options, which is less common for air-to-water systems.
- *Airflow*: the effectiveness of fan heating is reliant on air movement within the building. Understanding airflow in your home or business premises will help you get a better picture of the performance that you can expect from your heat pump, and which type is better suited to your home.
- *Hot water*: air-to-air cannot be used to meet hot water needs.

Is your property right for an ASHP?

- *Space*: you will need space outside (or inside if using heat recovery) to place the evaporator unit. They can either be fitted to a wall (a sunny wall is best) or placed on the ground, but crucially there must be enough space surrounding the unit to enable a good airflow.
- *Insulation*: ASHPs produce heat at lower temperatures than conventional heating systems. Therefore it is essential that your property is well insulated if your heat pump is to function efficiently. Otherwise much of the heat generated will be wasted in leaks, making the system less effective, more costly and inefficient. Remember that installing a heat pump in a poorly insulated building can actually lead to higher energy bills, since the emitters are inevitably warmer, which makes the COP lower.
- *Type of heating system*: ASHPs perform best when used with underfloor heating or fan systems, rather than conventional radiator systems, although low-temperature radiators can be used if designed properly.
- *Timing*: combining the installation with the development of a new-build or a renovation can decrease the costs significantly.

Buying and Installing

Generally speaking, there are certain conditions that make selecting a heat pump the right choice for heating or cooling needs:

- if the alternative energy source is expensive or unavailable (e.g. all electric heating can be replaced or where there is no mains gas);
- if the heat pump can deliver heat via underfloor heating (e.g. for new-build or refurbishment);

- if for preheating water or for only relatively small volumes of hot water will be needed; and
- if it is possible to power the heat pump from an onsite source of renewable electricity.

Particularly since heat pump systems tend to have relatively high capital costs, getting an accurate calculation of the building's heat loss is both important and the first step to take before selecting the best system for your needs. The related energy consumption profile and the domestic hot-water requirements also need to be considered in order to get the figures for an accurate sizing of the heat pump system. Oversizing a heat pump system will result in increased periods of operation under part load. Frequent cycling (automatically turning itself on and off) reduces efficiencies as well as equipment life. If the system is undersized, on the other hand, it is probable that top-up heating will be needed to supplement the system, increasing costs and reducing system efficiencies.

Since most heat pumps can operate in heating or (in reverse) cooling modes, the relative coefficiencies of both modes should be investigated before purchase. If you intend to use it for both heating and cooling, for optimum efficiency the model with the best coefficient of performance for your primary use (heating or cooling) should be selected. In the southern US, for example, the cooling mode should be prioritised, while in most of Canada the heating load would be more appropriate.

There are many companies selling and installing heat pump systems. Because the quality of the instalment itself will have a significant impact on the performance of the heat pump, it is important to do some research into the different options available so that you have a good idea of what you want from your system and from your installer. Some companies offer tours (site visits) for prospective customers around completed installations, which can be useful when trying to envision how a heat pump might work in your home. It can also give you the opportunity to ask plenty of pertinent questions.

There is a particular performance issue to bear in mind with ASHPs. According to a report on research funded by DOE's Energy Star programme, over 50 per cent of all ASHPs have significant problems with low airflow, leaky ducts and incorrect refrigerant charge. Refrigeration systems should be checked for leaks at installation and during each service, and it's important to ensure that the installing technician measures airflow before checking refrigerant charge, because refrigerant measurements are not accurate until the airflow is correct.

Table 5.2 Low temperature water-based distribution (e.g. underfloor) is the most compatible and efficient for use with heat pumps

Distribution system	Delivery temperature °F	Delivery temperature °C
Conventional radiators	140–194	60–90
Low-temperature radiators	113–131	45–55
Underfloor heating	86–113	30–45
Air heating	86–122	30–50

Source: Energy Saving Trust www.est.org.uk

The type of heat-distribution system is another important factor to consider. Heat pumps becomes less efficient the higher the temperature required by the heating system, so they are rarely used for heating hot water for purposes – such as for baths, showers or washing dishes – that are not directly related to space heating. Heat pumps can heat water up to 50°C or slightly more, but beyond a target temperature of about 35°C, the efficiency involved in achieving this level drops off fast. It's worth noting that water temperatures of 60°C or above are needed to prevent Legionella. It is entirely compatible to install an integrated solar water-heating system (see Chapter 4) to augment the temperatures of water pre-heated by a heat pump. For good environmental practice, it is important to select non-toxic circulating fluids for use in sealed (indirect) systems and open (direct) systems.

When choosing a unit and an installer it is recommended that you find a reputable installer and choose equipment that has been accredited by the relevant bodies. Where grants or incentives are available, one of the criteria for eligibility may be the use of a pump or installer accredited with bodies such as the ARI (Air Conditioning and Refrigeration Institute) in the US and the MSC (Microgeneration Certification Scheme) in the UK. See Annex 2 for further details.

6
Photovoltaics

There is more than one way to generate electricity from sunlight but, in practice, virtually all solar electricity generation in the world uses photovoltaic cells, also known as solar PV. Photovoltaic modules (or PV modules) are generally mounted in metal frames and installed in 'arrays'. Typically one or two m^2 (10 to 20 ft^2) per module, they can be wired together to create arrays of several kW (1,000 Watts) or even MW (1,000 kW). Apart from its dependency on the sun as the direct source of its energy, PV has little more in common with solar thermal collectors than that they are often found on the roofs of buildings, oriented towards the best possible view of the sun's path. PV modules generate electricity while solar collectors absorb heat.

Developed initially for powering spacecraft, satellites and other equipment in space, PV is seen as the most expensive way to generate electricity. Ironically, though, it's also the most likely to be used by the poorest and most remote communities. Indigenous villages in the middle of the Amazon, for instance, are impossibly far from grid connection, yet the transport, time and money needed to run diesel generators for powering essential radio communication, lighting or refrigeration systems would be much more expensive than PV. This factor, combined with the reality that PV modules are simple and pleasing items for NGOs to donate to needy communities, helps explain the irony.

PV costs are coming down fast, but in 2011 a domestic scale grid-connected PV system (in the 1 kW to 6 kW range) cost between about US$6,000 and US$30,000, depending on kW output and the quality and sophistication of the modules, installation and controls. This relatively high cost, however, has not stopped PV being introduced into North America and Europe over the last ten years. There are two main reasons for this: green vision (from businesses and households for credibility, PR and/or genuine environmental commitment) or financial incentive (usually from government, e.g. the feed-in tariffs FiTs of Germany and the UK, and rebates or capital incentives from US states such as California and Illinois).

In recent years, incentive has taken over from vision as the main motivating factor in many places. At the time of writing, the preferential feed-in tariffs in the UK make investing in PV worthy of any business portfolio, let alone a household hoping to economise. Capital costs are likely to be paid off in a time-period of seven to ten years, after which the electricity used in an approved PV powered building is essentially free, with any surplus being sold into the grid at very good rates. In addition, maintenance costs are very low on grid-connected systems.

In remote situations, whether in the developed or developing world, the

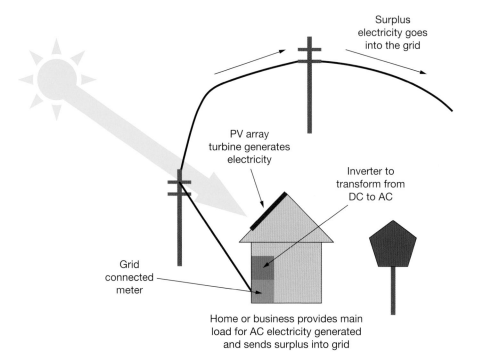

Figure 6.1 Typical grid-connect PV system serving domestic or business premises with surplus solar power being sold into the grid, and the grid providing power if not enough is being generated by the solar modules to cover building electricity demand

Source: Dilwyn Jenkins

argument for PV will generally depend on the cost of the alternative source of power, be that mains connection or local generator. In this chapter we focus mainly on the grid-connect situation, which is that most likely to be encountered by householders and business managers in the English-speaking world. There is also a brief section below on off-grid PV.

Basic principles

A number of photovoltaic cells, usually made of silicon, are mounted together in a module or panel. These modules are connected together to produce the amount of power required. They can generate some electricity even when they are not in direct sunlight, or if the light is diffused by cloud cover.

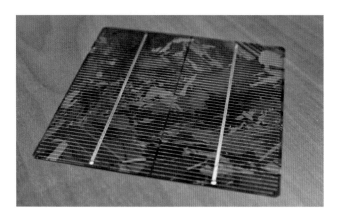

Figure 6.2 Photovoltaic cell (polycrystalline type)

Source: Dulas Ltd, www.dulas.org.uk

PHOTOVOLTAICS 113

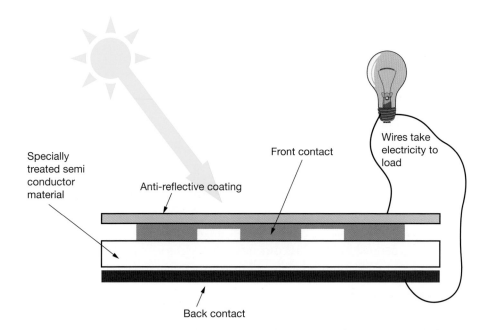

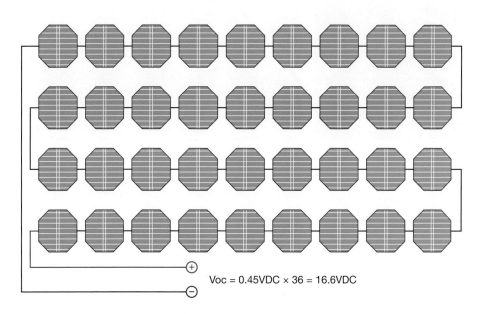

Figure 6.3a Diagram showing cross-section of photovoltaic module and **6.3b** photovoltaic module showing how the solar cells are connected in series

Source: (a) Dilwyn Jenkins

PV cells produce direct current (DC): the type of electricity stored and delivered by batteries. Most electrical equipment runs off alternating current (AC), as supplied by most electricity grids throughout the world. So a PV system usually needs to be augmented by an inverter: an electronic device that converts DC electricity into AC.

PV has a number of advantages over other forms of electricity generation, since it:

- requires no fuel to operate;
- produces no CO_2 emissions during operation;
- is silent in operation;
- is highly portable;
- can be building-integrated or racked;
- is highly modular (more modules can be added later if space is left);
- has no moving parts;
- can operate for 25 years or more with virtually no maintenance; and
- requires no water for cooling, in contrast to coal, gas and nuclear power stations.

PV, however, still has the one big disadvantage of being relatively expensive. Electricity generated by PV cells costs more than electricity generated by any of the other technologies in this book, and by conventional fossil-fuel technologies

Figure 6.4a Close-up photograph of Kyocera polycrystalline PV module

Source: Dulas Ltd, www.dulas.org.uk

Figure 6.4b Photograph of polycrystalline PV module fitted into larger array

Source: BSW-Solar Viessmann, www.solarwirtschaft.de

Figure 6.4c Photograph of grid-connected polycrystalline PV module array on roof of business units with single module PV car-park lighting in foreground

Source: Dilwyn Jenkins

Figure 6.4d A PV array with a solar water-heating system installed below it

Source: Frank Jackson

supplying the electricity grids of the world. For this reason, PV has three main markets: regions that have attractive feed-in tariffs or net metering for PV (if a government wants to support the widespread uptake of PV, usually because of the CO_2 benefits, then the preferred way to provide support is through offering a higher price for PV-generated electricity than for other sources); remote areas where there is no mains electricity, and where running small-scale diesel generators would be too expensive or inappropriate; and very low power loads, where it is not worth providing any other power supply given the small PV requirement (e.g. motorway road signs, garden lights and calculators).

In countries where preferential tariffs or net metering are available, PV installations are becoming common on houses and commercial buildings, supplying electricity for use onsite and selling any surplus to electricity companies for use elsewhere. If the economics stack up, as they presently do in many parts of the US, the UK and other English-speaking countries, large-scale PV farms are also installed, selling all their output directly into the grid like any other power station.

How Much Electricity do PV Modules Generate?

The amount of electricity generated will depend on a number of factors, including the area of a PV array, the efficiency of the cells and the amount of sunlight hitting the cells. Because of these variables, PV modules are always described by quoting their peak output, in watts peak or kilowatts peak (Wp or kWp). Module outputs increase with the size of the module and the commonest fall into the 5 W to 200 W peak output range. The peak output of a module is the amount of electricity it can theoretically produce in direct sunlight with an intensity of 1,000 W/m^2 at 25°C. Most sites in the world will seldom get sunlight brighter than this, so the peak output is effectively the maximum electrical output you can expect in perfect weather and temperature conditions.

The module type and location in the world are the two most important factors for determining the amount of energy that can be produced in a given year. The rated power of a module (calculated by the manufacturer) in watts (Wp) can be used to work out the annual energy output of a module or array of modules in relation to the insolation data (solar radiation data) for the location. This data comes as kWh/m^2 per year. So, assuming a module rated at 150 Wp and insolation data of 900 kWh/m^2 per year, also referred to as 900 peak sun hours per year, the annual output calculation would be: 150 W $\times$ 900 $\times$ 0.75 (performance ratio) = 101,250 Wh (or 101.25 kWh) approximately annually.

The 'performance ratio' is an approximate figure to account for things such as a module's lower output in high operating-temperatures and system losses. If you know the peak output and you know how sunny the site is, and a roof is close to the optimum angle and orientation, then it's possible to make a rough estimate as to how much electricity you will get from your modules. You don't need to know how efficient the cells are, what the effective area is or what type of cells you have in order to do this calculation. There are some user-friendly online calculators (see below), however, you might also want to know how quickly the peak output will drop off over time and how efficient the inverter is. Furthermore, there is the issue of shading from the sun (more on shading, and on estimating the output of a particular system, in Buying a PV System below).

How do PV Modules Work?

In essence, a solar cell or a photovoltaic cell is a flat semiconductor device that converts sunlight into direct current (DC) electricity. They are made from thinly sliced baked silicon wafers. If you attach a circuit to both sides of the cell, electrons will flow through the circuit from one side of the cell to the other, and in the process a very clean form of direct current electricity is generated. PV modules are made by combining a number of solar cells into one module. Then, at the solar system's design and installation stages, modules can be combined together to make a solar array of any required size and output.

Generally speaking, each silicon wafer is around 10 cm (or 4 inches) wide and, as a diode, produces roughly 0.6 V DC. At an atomic level, silicon transmits energy to an electron on its outer surface when struck by a photon (the quantum particle of sunlight). The two-layered structure of the solar cell makes it possible to capture the energy and encourage it to flow in a circuit. The top silicon layer is injected with phosphorous, while the lower layer with boron (providing, respectively, the negative and positive potential). With 36 cells working together, the total theoretical peak output (known as the 'open circuit' rating) would be 21.6 V DC. However, modules can be configured to produce a wide range of voltages. The current produced by the cells (or the modules or array) depends on the surface area and the level of solar insolation. Voltage multiplied by current gives the power produced ($P = VI$).

Types of Photovoltaic Modules

There are four main types of module on the market: crystalline silicon PV is the most commonly used (either polycrystalline or monocrystalline); thin film

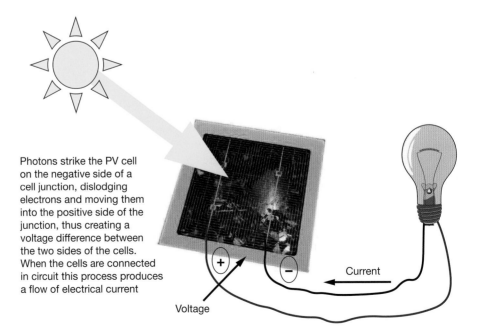

Figure 6.5 A solar cell at work

Source: Dilwyn Jenkins

modules; and the hybrid module type. Most PV modules are guaranteed to still produce 80 per cent of the rated electricity (compared to output when brand new) after 20 to 25 years of service

Monocrystalline Silicon Modules

Figure 6.6a (left) Monocrystalline cell

Source: ErSol Solarstrom GmbH & Co. KG, www.bosch-solarenergy.de

Figure 6.6b (above) Monocrystalline module, note the distinctive hexagonal shape of the cells

Source: Dulas Ltd, www.dulas.org.uk

Monocrystalline modules are made from slices of a single grown crystal of pure silicon. The slices are cut into a distinctive shape: a square with the corners missing. Monocrystalline cells have been in use for many decades and have a proven long life, with efficiencies of 14 to 20 per cent.

Polycrystalline Silicon Modules

Polycrystalline modules (sometimes called multicrystalline) are made from larger slices made up of multiple silicon crystals, often melted and cast. Less expensive than monocrystalline to manufacture, they have a distinctive 'crazy paving' pattern and no visible gaps between the cells (see Figure 6.4a). They have a proven long life, but efficiency is slightly lower than monocrystalline cells, at 13 to 15 per cent.

Thin-Film Modules

There are three types of thin-film modules currently available on the market: amorphous silicon, cadmium telluride (CdTe) and copper indium selenide (CIS)/

Figure 6.7 Uni-Solar's thin-film flexible solar modules being installed on a roof

Source Ken Fields, Wiki Commons, http://commons.wikimedia.org

gallium diselenide (CIGS). Thin-film modules have a uniform, very thin, layer of silicon sprayed onto a sheet of glass or a similar material. They usually have a continuous and even, if oily, colour apart from a pattern of parallel lines created by the electrical connections. Life expectancy has been improved over the years, but is not as well proven as for crystalline silicon. Efficiencies are considerably lower than crystalline, usually less than 12 per cent or lower, and these modules are not used so widely in general power production for homes or businesses.

There are other materials that can be used to make PV modules. Gallium arsenide cells can have efficiencies greater than 30 per cent for example. However, all these other technologies are either still in development or simply too expensive for normal terrestrial use.

Hybrid

Hybrid modules use a combination of crystalline and amorphous silicon, and they achieve the highest efficiency of all four types, together with a high life-expectancy. When first introduced they were more expensive per kWp than the alternatives, but the price differential is now often negligible.

Figure 6.8 In this hybrid (NA-F135) module the tandem structure is made up of an amorphous and a microcrystalline silicon layer

Source: Sharp Electronics (UK) Ltd, www.sharp.eu

Relative Efficiencies of Module Types

In order of efficiency, hybrid is the best, followed by monocrystalline and then polycrystalline, with amorphous trailing behind. However, you should never use efficiency as the sole consideration in choosing panels. A more efficient 2 kWp system will not produce any more electricity than a less efficient 2 kWp system, it will just have a smaller surface area, and if it costs more, is it worth it just to save some space?

Variations on the Standard Modules

Apart from the cells themselves, there are a number of variations that people have developed to try and get the most out of this expensive technology. The number of modules on the market is now very numerous, and they include PV tiles, slates and various productions for building integrated PV (BIPV). One example is the hybrid solar-thermal PV panel, or PVT. This has a fairly standard PV module with a flat-plate solar thermal collector beneath (see Chapter 4). The solar thermal collector takes heat away from the PV cells, so increasing their efficiency (PV is more efficient when operating at lower temperatures).

Figure 6.9 A hybrid system where electricity and hot water are produced simultaneously. This manufacturer produces two versions, one of which is designed to optimise the electrical output while the other optimises hot water output

Source: Solimpeks Solar Energy Corp., www.solimpeks.com

However, there will always be a compromise between electrical and thermal performance. How PVT panels perform in practice remains to be seen, and whether they are worthwhile financially will depend on how the support mechanisms in your country choose to deal with it. Concentrated PV (CPV) is another alternative. In CPV, lenses are used to magnify the power of the sunlight, but these are used almost exclusively on solar farms in regions where levels of direct sunlight are high.

Module Mounting Systems

The main methods of mounting PV modules, relevant for homes and businesses, are:

- *On-roof mounting* (on sloping roofs): these are usually on prefabricated rails or mounting brackets, both of which are fixed through the roof and roofing material onto roof joists below. The PV modules are then fixed, usually with bolts (sometimes clips) to the rails or brackets.
- *Flat-roof mounting*: these generally use metal frames that can be angled according to the requirement of the latitude in the specific location. They are usually weighed down with stone, gravel or concrete ballast and incorporate integrated cable channels. In regions with very hot summers, the roof mounted PVs can help to keep a building cooler.
- *Ground-mounting*: similar to flat-roof mounting except that the module-mounting structure is on the ground. They usually have metal frames, which can be angled to the requirements of the specific latitude and are fixed to concrete foundations.
- *Building integrated PV (BIPV)*: where the modules are integrated into the building structure (although sometimes roof mounted systems are included in this term, even though they are only bolted to a building rather than integrated into the structure itself). Simultaneously serving as building envelope material and power generator, BIPV systems can provide savings in materials, electricity costs and GHG emissions. They are generally integrated into glass on the facade of a building, and are currently most common in commercial premises, where they can also assist with shading from solar glare (e.g. on computer screens in offices).

Figure 6.10 On-roof mounted PV array in the UK (in this instance the house is grid-linked and sells surplus electricity into the grid, benefitting from preferential feed-in tariffs)

Source: Dulas Ltd, www.dulas.org.uk

Figure 6.11 Flat-roof-mounted system: The Energy Roof at Sheffield Gazeley Blade, UK

Source: Solarcentury Ltd, www.solarcentury.co.uk

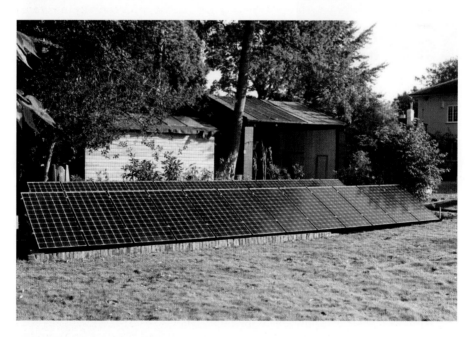

Figure 6.12 Ground-mounted PV modules in system owner's garden, Kent, UK, from the *Earthscan Expert Series* book *Grid-connected Solar Electric Systems*

Source: Paul Barwell

Figure 6.13a, 6.13b and 6.13c Building integrated PV at commercial offices, Doxford, UK

Source: Schuco International KG
www.schueco.com

Common PV Installations

PV is used for a wide variety of applications throughout the world, but for most householders and businesses there are a few key models that will cover most installations.

The Grid-connected House

Usually between 2 and 4 kWp of modules are fitted to the roof of the property, either directly to the part of the sloping roof that is closest to south facing, or on angled mounting frames on a flat roof (south of the tropics the panels will face north.) The modules are connected together and the output fed into a (usually) single inverter. The inverter produces mains voltage AC electricity that is compatible with, and meets the standards required for, mains electricity in that country.

The output from the inverter is fed into the house electricity-supply circuit, and if the house is using more electricity than the panels are producing then all the solar output is used directly by the household. When the PV output exceeds household consumption, the excess electricity flows back into the grid along the existing supply wires and is used elsewhere by other consumers.

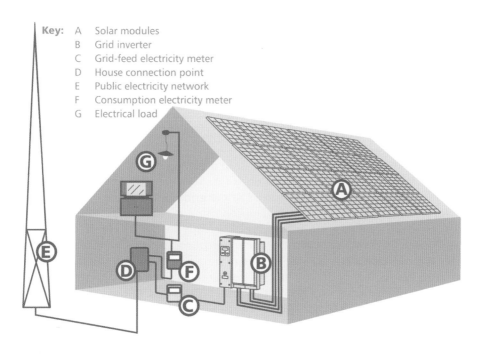

Key:
- A Solar modules
- B Grid inverter
- C Grid-feed electricity meter
- D House connection point
- E Public electricity network
- F Consumption electricity meter
- G Electrical load

Figure 6.14 The components of a grid-connected system. The modules feed power to the home's appliances through the grid inverter, which produces grid-quality AC electricity. Any surplus is sent to the local electricity network through the grid-feed electricity meter. When the home's demand is greater than can be satisfied by the PV array, such as at night-time, power is drawn instead from the local network through the consumption electricity meter. The Utility Company would bill the home for the balance of the amount of power used and supplied, according to the tariffs agreed for the electricity bought and sold.

Source: Steca, www.steca.com

You need approval from the electricity supply company or network operator to connect to the grid. In some countries this has been made as simple as possible, and companies may be required to permit connection and to buy the excess electricity, provided that you meet certain criteria. In other countries it may not be so straightforward but may still be possible. If it is possible, this is generally the best way to make use of your surplus electricity. Batteries are not normally used with grid-connected systems.

Figure 6.15 A grid-connected house with solar thermal and PV on the same roof

Source: BSW-Solar Viessmann, www.solarwirtschaft.de

You will usually need additional meters to benefit from any support for generating renewable electricity, but the requirements vary in different parts of the world. In principle, there are two types of metering systems: net metering and gross metering. Net metering is when property owners are allowed to produce enough electricity to cover their own needs only, while with gross metering all electricity can be sold onto the grid. However, the details of these arrangements differ considerably from place to place, and there are many variations. Net metering tends to be more common in the US, while in Europe many countries have gross-metering schemes.

For example, in the UK you will need a total generation meter after the inverter, as householders are paid a tariff for every unit of electricity generated, whether it is used on site or exported to the grid. Householders are also paid extra for electricity exported, but the proportion exported is usually estimated and so an additional export meter is not currently required in the UK. All electricity generated and used by the household reduces electricity imports, and this saving shows up on the existing electricity supply meter. Fuses and other safety measures as required by local codes are also needed, and the generation system needs to automatically shut down if the grid fails so that workers can repair the line without risk of electrocution.

In temperate countries, a domestic PV system of this type might typically produce enough electricity to offset about half of the electricity used by an average household, although energy efficiency measures can increase this considerably. Until recently, grid-connected domestic PV systems were only installed by relatively well-off, environmentally concerned people who wanted to do their bit to combat global warming. However, since the introduction of net metering and feed-in tariffs in some countries, many people are investing in a PV system primarily for the secure financial payback it offers. It may not necessarily pay to borrow money to put PV on your roof, but if you have money sitting in a savings account earning next to nothing you could be better off investing it in silicon.

Business and Other Non-domestic Users

A small non-domestic system may be very similar to a domestic system, with the advantage that peak electricity demand is more likely to coincide with peak solar output during the day. Legislation, support and tax issues may be different for non-domestic generators.

If the company's electricity supply capacity, or the PV system's peak output, are larger than a domestic system, then more stringent connection requirements may kick in, requiring more extensive safety and quality-control measures, increased installation costs and possibly longer approval times. In particular, a three-phase connection may be required, especially if the business already has three-phase supply. This will require three inverters, and three times as many meters, adding to the cost.

In a similar way to householders, the only companies that used to put PV on their offices were those with a very specific green marketing strategy and the budget to support it. Feed-in tariffs and net metering have changed that, and companies now regard PV as a reasonable investment option in its own right.

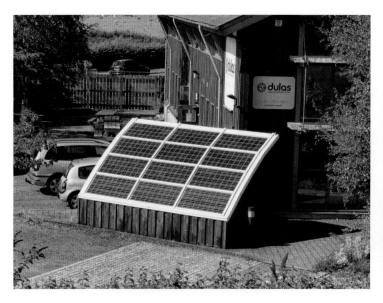

Figure 6.16a PV array, installed in the late 1990s and designed to meet 10% of the building's electricity demand, at Dulas office buildings, Wales, UK

Source: Dulas Ltd., www.dulas.org.uk

Figure 6.16b Recently completed PV grid-connect installation on roof of office building in Stroud, UK

Source: Severn Wye Energy Agency, www.swea.co.uk

Off-grid Systems

Off-grid PV systems, or stand-alone PV systems as they are also called, are PV systems that operate independently of the utility grid, usually storing the electricity generated in a battery bank. They can be used to run lights and essential appliances for an isolated home, a fridge for a mobile medical unit, a telecommunications transmitter on a mountain-top, or anything that needs a small amount of power and doesn't have grid electricity close by.

The output of the PV modules is fed, via a controller, to a bank of batteries. Low voltage DC lights and other appliances are connected via the controller, and take their power directly from the PVs or from the batteries. The lights, appliances and batteries all work on DC, just like the PV modules, so no inverter is needed. If well maintained, good quality batteries will usually need replacing every five years or so.

Larger systems, and those operating specialist equipment, may also have an inverter to supply AC, and they may have a diesel generator and perhaps another renewable generator such as a wind turbine to improve reliability or total output. These other generators can all be used to charge the batteries as required, but the diesel may also be used to run occasional high power-loads directly.

Figure 6.17 Diagram of an off-grid PV system

Source: Steca
www.steca.com

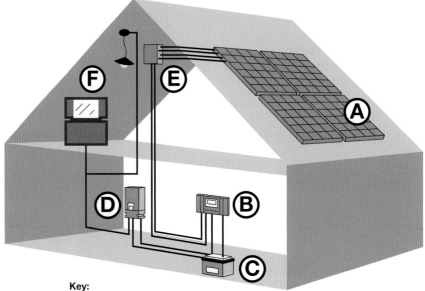

Key:

A Solar modules
B Solar charge controller
C Battery
D Sine wave inverter
E Generator junction box
F Electrical load
  (12 V ... 48 V DC, 115 ... 230 V AC)

Buying a PV System

All PV modules are sold labelled with their peak output. This is measured according to internationally agreed standards, and you can generally rely on the stated peak output in comparing one module with another. By shopping around it is possible to compare the costs per kW output of the different modules and makes available. To do this, simply divide the cost of the module by its rated output in Watts: $500 per module that has a rated output of 100 W = $5 per Watt

If you want to make your own estimate of likely output there are a number of online tools to help. You will need to enter your location, the angle and orientation of the panels, and the peak output of the array. Usually you can either enter details of the modules and inverter or use the default settings.

As a rough guide, a well-oriented unshaded system in Washington State, US or in the UK, will probably produce about 850 kWh per year per kWp installed. The same system in sunny Arizona would produce more like 1,500 kWh per year. There are some easy-to-use PV system-sizing software programmes available online:

- PVGIS – www.re.jrc.ec.europa.eu/pvgis/
- PVWatts – www.nrel.gov/rredc/pvwatts/
- SMA iPhone app – www.sma-america.com/en_US/products/software/sma-solarchecker.html

If space is limited and you want the maximum possible output you should choose the most efficient affordable modules, which at the moment are likely to be hybrid modules. You will pay more because you are buying more kilowatts: the main module types cost roughly the same per kWp, so the more efficient types cost more per square metre.

Systems need to be installed well. PV modules are guaranteed for 20 years or more, so the installation should be designed to last at least that long. It's also important to insure an investment in a PV system, so that if anything unforeseen happens, you are covered for repairs or replacement. The main areas where a PV system might be poorly installed include the mounting system and how it is fixed to a roof or frame, the wiring layout (e.g. array conductors should be neatly and professionally in place), and safety issues, such as proper module and array grounding to earth, and protection from lightning. Advice on insuring a PV system can be obtained from solar trade-associations and the relevant government agencies (or their websites).

These days, the sizing of most grid-connect PV systems will depend largely on the incentive system in place, whether this is net metering or a feed-in tariff. It will also depend on the levels of solar insolation your area receives over a year and at which points in the year. If the incentives are sufficient for even a large array to pay back within five to eight years, then profit from the sale of green electricity to the grid could be the main motivating factor. As such, the larger the system the better, and it should be possible to get a bank loan for the project.

In a net-metering situation, most PV users are going to want to at least cover their own electricity needs. When sizing a PV array, calculating exactly what your electricity needs are should be one of the first steps, but given the high capital cost of PV, some businesses tend to choose a percentage of their electricity needs to be covered by PV, as a green statement and contribution towards emissions reductions. In the Dulas example (see Figure 6.16a) this was 10 per cent, back in the time that this was a UK government target for emissions reductions. These days the targets are necessarily higher.

There are few essential considerations to bear in mind before buying PV, whether it's on or off-grid:

- How good is your solar resource?
- How good is the local solar aspect?
- Is there shading (if caused by trees this might be possible to rectify, but with buildings this is obviously more difficult)?
- Is the system to be building integrated, on racking or on a tracking rack?
- Are there other environmental factors (like snow or ice build-up in winter, dust in summer, salt in the air if near the coast)?
- Make sure that the modules have sufficient airflows to stop overheating in particularly hot environments.

Whichever PV system type is chosen, it is important to have a professional survey carried out by a PV installer–supplier. When working out where to put your modules, and what they will produce, your installer will need to consider many aspects very carefully, from how to fix the modules to issues like shading. If just one corner of a module is shaded it can seriously reduce the output of the whole module, and output can also depend on how the modules are wired together. The ideal is to arrange the panels so that none of them are ever shaded, but that isn't always possible. If some shading is going to happen at certain times of day then there are things that can be done to minimise losses, such as having several inverters or microinverters, but this is not straightforward and mistakes can be costly, so professional advice should always be sought.

Grid-connected photovoltaic systems are not suitable for DIY installation because of a number of electrical safety issues and the importance of quality. The systems also have to meet strict technical criteria to be legal. Using an approved installer is the easiest way to achieve this and may even be a legislative requirement for capital grants or preferential tariff-rates since financial support is often only available for systems commissioned by approved installers. Off-grid systems tend to have fewer legal restrictions, but even a competent electrician may need specialist help to install a system that will operate effectively without damaging the batteries.

Countries that provide financial support for PV systems will have one or more accreditation schemes for installers, and this will be the easiest way to find an installer. For example, in the US the minimum requirement would be a C-10 (Electrical) licensed contractor, but better still is certification from the North American Board of Certified Energy Practitioners (NABCEP) and Energy Star Registration (under Solar Energy). In the UK, all approved installers are members of the Microgeneration Certification Scheme and can be found on the MCS website. Usually the modules, and other equipment, will be supplied by the approved installer. You may choose to source these things yourself, but you will then need to make sure that all the equipment meets any requirements that are needed to make them eligible for support. The approved installer is usually in a better position to do this.

It is always worth trying to get recommendations of good installers from satisfied customers. If you don't know any existing customers, ask installers to give some reference installations. It is also sensible to get more than one quote. The price from different installers will vary far more than the price of different cell technologies, and you can save thousands on even a simple domestic system by shopping around. This is particularly true in countries where PV systems are still uncommon and there is little competition. This means that the harder it is to find three companies to quote, the more you are likely to benefit from getting three quotes.

Costs and Payback

PV module production is a global business, and the wholesale cost of modules can be quoted as a single statistic. However, the cost of buying modules, the inverter and the meters, and of having it all delivered and installed at a particular location can vary considerably. The bigger and more established the PV market is in a country, the cheaper the installed cost-per-kWp. Wherever a PV array is to be installed, the expense, life expectancy and safety issues associated with the system all warrant the use of professional advice from the beginning and experienced, qualified and certified installers.

In 2009 a domestic PV installation would typically have cost from around US$7,000 per kWp in the US and around €6,000 per kWp in the UK. After the feed-in tariff was introduced in the UK, installation numbers shot up to 25,000 in the first year, and by 2011 the price had dropped to about €4,000 per kWp. In Germany, where feed-in tariffs have been available for several years, the price is below €3,000 per kWp. Larger systems can be installed for even less. This is still too expensive for systems to be economically attractive without government or utility support, but it does mean that, as overall costs approach grid-parity, the extent of that support is reducing rather rapidly. Grid parity is when the cost of PV generated electricity is the same as that bought from the grid by the end user.

The main capital costs lie in the equipment – PV modules, inverters etc. – which account for around 80 per cent of the total expenditure on average. In most areas of the US, utilities are obliged to offer at least free net-metering and a free meter. This helps the payback scenarios, as discussed below. Operation and maintenance costs for PV systems are negligible (perhaps 1 per cent or less of the overall lifetime system cost). Inverters are usually covered by warrantees of 5 to 10 years and are unlikely to last as long as the panels, so you should budget for the possibility of replacing it at least once over the lifetime of the system, although some inverter manufacturers do offer longer warranties.

Financial Support and Feed-in Tariffs (FiTs)

Many governments and other organisations have offered capital grants for PV systems over the years in an attempt to encourage greater uptake. A grant of well over 50 per cent would be required to create an attractive opportunity for investors. Schemes that offered grant support of up to 50 per cent have encouraged small numbers of installations in niche markets, but they have all failed to bring the sort of mass uptake that would lead to cost reductions and hence an affordable technology.

Feed-in tariffs (FiTs) offer a different approach to providing financial support which, for PV in particular, has proved considerably more successful. There are two main forms of FiTs: the gross FiT (where the generator gets paid for all the electricity generated, even if some has been used by the house or business property generating) and the net FiT (where only the electricity actually exported to the grid is eligible). The broad principle, however, is that electricity supply companies, even if they are independent commercial companies, are required to buy electricity generated from private PV installations at a premium price: the feed-in tariff. In practice, because PV generation brings pretty much the same CO_2 benefits whoever uses the electricity, generators are often paid a tariff for every unit they generate, even if they don't sell the electricity itself.

The most relevant features associated with FiTs are the rate offered and the period or lifetime of the offer. If the rate is good for a 25-year period, there is significant incentive to invest. To give certainty to investors, successful FiT schemes include a guaranteed level of support over an extended lifetime, often 20 or 25 years, and often index linked to maintain the tariff's value.

This system has a number of advantages over grant schemes. The advantages from the government's perspective are that:

- There is no upfront payment required: the private sector provides all the capital, and the cost to government or electricity industry only arises as the benefits are realised.
- Tariffs can be adjusted to avoid over-compensating once installation costs have reduced.
- The onus is on the installer/supplier to install efficient systems: a badly-sited PV array will generate little electricity and so will generate less income.

The advantages from the system owner's perspective are that:

- Because of the above, the government can afford to give adequate support, even for early adopters, and so PV installations become an attractive investment.
- The guaranteed future tariffs give confidence that this is a sound investment, both for those investing in their own scheme and for lenders supporting projects elsewhere.

Germany was the first country to implement a successful feed-in tariff that adequately compensated small generators, including householders and small businesses. To date, nearly 30 GW (30,000,000 kWp) of PV has been installed in Germany; in California, the primary user of PV power in the US, the total installed capacity in 2010 was just under 1 GW, but the State is expecting to install a further 7.5 GW by the end of 2015: enough to power around 2 million Californian homes. Other countries have followed suit. In the UK 25,000 homes installed PV systems in the first year of the scheme, representing about 60 MW (60,000 kWp).

The PV industry would like to see grid parity, where the cost of PV generated electricity is the same as that bought from the grid by the end user. There are, however, a lot of factors involved: system cost, solar radiation in the area and utility prices in the district. Germany hopes to reach grid parity for PV by the end of 2016, but is aided in this by high electricity prices and a competitive industry.

7
Wind Energy

Wind power was already harnessed by people from as far afield as the Mediterranean and China by the time Rome invaded ancient Britain. Yet only in the last 20 years or so has a really competitive market begun to appear for this technology that provides free and clean energy, if unpredictable and intermittent in its raw state. However, it doesn't really matter if the wind comes and goes as it pleases, we can use national electricity grids to 'store' and transport wind electricity so that it can be used when and where it's needed. For off-grid wind turbines, the technical fix is a battery store.

After decades of pioneering turbines, the large-scale wind industry took off in the early 1990s. This is still booming. Somewhat slower to emerge, but definitely now here, is a wide range of small- to medium-scale turbines, suitable for installation in homes and businesses. With the emergence of feed-in tariffs and supply payback schemes, the economics are significantly improved and the range is likely to continue growing. While wind power through the grid is appropriate, given the economies of scale offered by present wind technology one turbine for every single home or business is unlikely to ever happen. Most urban buildings simply don't have a sufficient or good enough quality wind resource in their immediate environment.

The size of the wind turbine needed depends entirely on the intended application. One fundamental issue is capturing enough wind and generating sufficient electricity to make it worth the investment: a payback concept. However, payback is neither essential nor likely to be the only motivating factor. Having a green conscience might be enough, using wind to attempt to reduce the carbon footprint of a home or business. You may only want to make a contribution to your electricity use. Particularly as a business, you might want it to impress people, such as clients in a relevant industry. There are plenty of possible reasons for investing in wind power, but the main one is to cover all, or part, of one's electricity needs, ideally at a lower cost than buying from the grid. In some cases this will extend to selling surplus electricity into the grid.

If wind power is an option for a home or business, the first essential task is to work out if your position has sufficient wind resource. Even if there are strong continuous winds, this resource needs to be assessed against the nearby terrain and any constructions or vegetation that might impact on the local wind-flow patterns. The next task is to think about whether you can achieve this without annoying the neighbours. The intended turbine site, whether it is building-integrated or free standing, will need sufficient space, particularly with regard to neighbours who might not feel the same way about wind power or who may object on the grounds of noise or visual impacts.

Given a reasonably good annual wind-regime, plus some open space near the property for a tower, it may be worth investigating further. More and more business units can be seen alongside the highways of Europe and North America with a small- to medium-sized wind turbine proudly standing high on its tower close by. In rural areas, too, small-scale turbines – from 1 kW to around 10 kW – can be seen with greater frequency, mainly associated with farm buildings.

To have any chance of making a serious contribution to a property's electricity demand, the average house (US properties tend to have higher electricity consumption than those in the UK, but the range of domestic electricity use is from about 3,500 kWh a year to about 10,000 kWh a year) would need a system of between 1 and 6 kW, depending on levels of energy efficiency implemented, plus the wind resource at the site and other location factors. A larger building – for instance a public or business premises – is more likely to require a larger turbine, perhaps in the 10 to 30 kW range. Smaller turbines usually generate between 50 and 500 W, being designed mainly for battery charging in the leisure market (e.g. for boats and caravans). There is also a range of small turbines suitable for mounting on buildings and some, including vertical axis ones, that are considered appropriate for the urban environment.

Horizontal and Vertical Axis Wind Turbines

There are two basic types of modern turbine: horizontal-axis and vertical-axis varieties. There are upwind and downwind horizontal axis turbines. Each has its specific advantages and disadvantages: neither type is likely to produce more power or last longer. Horizontal axis turbines usually utilise three, but sometimes just two, rotor blades. A few of the smaller horizontal axis turbines have more than three blades. Factors like outputs and longevity need to be compared between different available turbine options. Two of the largest manufacturers of small turbines – Bergey and Southwest Wind Power – both produce upwind machines. Southwest Wind Power also sell a downwind model, the Skystream, aimed at the domestic and business grid-tied market.

On an upwind model – the most common of the smaller turbines – the rotor is located at the front of the device and uses a yaw mechanism (a turbine's system for searching out the best wind direction at any moment, often as simple as a tail fin) to keep it facing into the wind. With upwind turbines there is less likelihood than with downwind ones that towers will interfere with the wind and cause some turbulence and power loss. However, upwind machines will need an extended nacelle (the main body of turbine, where the generator is usually located) to position the rotor far enough away from the tower to avoid any blade-strike problems. Furthermore, upwind blades are usually more rigid to avoid bending back into the tower, but this causes other stresses at the rotor hub.

Downwind turbines have their rotors on the other side of the turbine and their nacelle is often designed to seek the wind, negating any necessity for a dedicated yaw. Downwind rotor blades can be flexible, making them cheaper and also producing less rotor-hub stresses. Some downwind turbines (e.g. Proven turbines manufactured in Scotland) have hinges to allow the blades to flex back in very

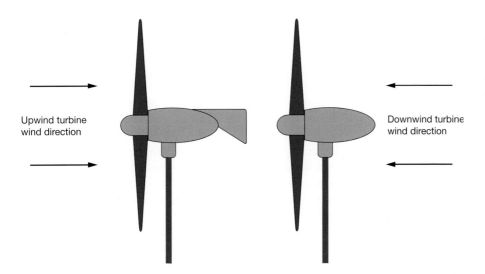

Figure 7.1 Upwind and downwind turbines showing different wind orientations

Source: Dilwyn Jenkins

high winds, dissipating some energy to gain speed control. Flexible blades are more prone to fatigue than fixed blades.

Nearly all smaller wind generators operate at variable speeds and were originally designed for stand-alone installations to generate electricity for a battery-charging system, and very occasionally for heat. Small wind-generators generate AC current initially but the voltage and frequency varies with the wind speed. The AC current is rectified to produce DC, making them suitable for charging batteries. Some of the smaller, and all of the larger, turbines incorporate an automatic speed-governing system or device to ensure that the rotor does not spin too fast in very high winds and overload. These days, even small generators can also be connected to the grid. Small- to medium-sized wind generators can also be grid-connected or used in stand-alone set ups with battery-charging and

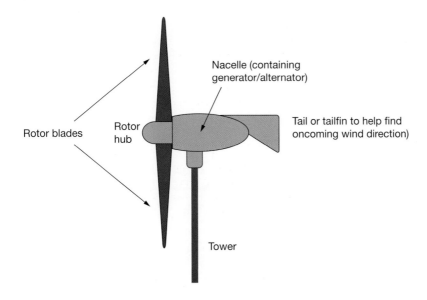

Figure 7.2 Main components of a small wind generator

Source: Dilwyn Jenkins

inverter systems. They are sometimes used very effectively in hybrid energy systems with a combination of diesel and/or solar PV inputs. Large-scale wind generators are almost always grid-tied and located in clusters or wind farms both on- and off-shore. They operate at a constant speed and are usually computer controlled. The larger turbines are now rated at over 2 MW each and have towers of 100 m or higher. The installed capacity of wind globally rose from 48 GW in 2006 to over 120 GW by the end of 2008.

Vertical axis wind generators – still generally named after Darrieus, their French inventor – have recently been incorporated into architectural projects. Darrieus-type turbines are shaped like an egg beater, and there are several new types of vertical axis generators which manufacturers claim are quiet, efficient, economical and ideal for residential energy production, even in urban environments. Vertical axis wind turbines are usually developed for urban situations, since it is here that most obstructions to smooth wind flows are found, and turbulence and changes in wind direction have fewer negative effects on vertical axis generators which are designed not to have to be positioned directly into the wind.

Savonius-type turbines use the wind to push the blades, so the rotation speed is always lower than the wind speed. The opposite effect is found with Darrieus-type generators, where the rotor can spin faster than the actual wind speed. Arguments abound comparing the efficiencies of these two basic types – horizontal axis and vertical axis. Horizontal-axis turbines work out cheaper per kWh produced, but vertical-axis turbines will work in almost any situation, including urban and rooftop, being less prone to serious impacts of local wind turbulence.

New vertical-axis type turbines are appearing on the market all the time, and there are now a wild variety of different designs and scales. This variety of technical options means that there is something suitable for most urban situations. However, it is still a developing technology and many are not yet beyond prototype stage. Building-mounted turbines have not had great market success yet, but

Figure 7.3 Vertical-axis turbine offsetting some power at a local government building, UK

Source: Dilwyn Jenkins

there are a number of models being tested on roofs and in corridors between buildings.

Grid-connected and Stand-alone Wind Systems

Wind systems are either stand-alone or grid-connected. The first is typical for generating in remote situations where wind is the best resource. Batteries are needed to store the power in stand-alone systems and the DC electricity stored can be connected to the grid via a suitable inverter. Small turbines with variable-speed devices will also need a power conditioning unit, or 'grid-tie inverter', to transform the current from 12 V or 24 V DC to the 110/230 V AC mains voltage. The Windy Boy from SMA is a popular inverter and there is a model certified (UL) for the North American market that offers automatic grid voltage detection allowing for straightforward and secure installation. US Federal regulations under the Public Utility Regulatory Policies Act (PURPA, 1978) require that electricity utilities connect to the grid and buy power from small-scale wind generators. With this in mind, the local utilities are able to advise on connection requirements for any new wind system destined for grid linking. In the UK, two-way meters are used, particularly in conjunction with the feed-in tariffs (FiTs) scheme, which pays preferential rates for clean electricity. Since none of us necessarily use electricity at the same time that the wind decides to blow, even with a grid-connected wind turbine that can theoretically cover all your power needs there are likely to be times when you will have to use the grid as a backup.

Heating with Wind Power

Much less common is the practice of heating space or water from wind energy. Water heating is frequently used as a 'dump' of surplus electricity in stand-alone systems, but the use of wind electricity for space heating in winter requires a

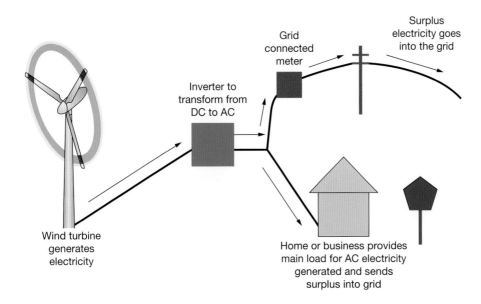

Figure 7.4 Diagram of typical grid-connect wind turbine set-up

Source: Dilwyn Jenkins

dedicated turbine. Keeping warm should not have to depend on how windy it is, even if there is some limited correlation. A wind heating system is only likely to be considered as a last resort (i.e. where there's no wood available and little opportunity for passive solar). More logical and efficient is to use a wind turbine to power a heat pump (see Chapter 5). This way it is possible to produce up to three times as much heat from the same amount of electricity generation.

Wind Energy Physics

Wind really is a free source of solar energy, created by the sun's uneven heating of the atmosphere. The rotation of the planet, irregularities in the surface of the earth – oceans, vegetation, buildings and topography – adds some specifics relating to how wind behaves at any particular site into the mix.

Wind energy devices are designed to generate mechanical power by converting the kinetic energy in the wind into mechanical power. This power can then be used for specific tasks: typically milling and pumping. Alternatively, a generator can convert this mechanical power into electricity. The electricity itself is made when the blades, spun by the wind, turn a copper coil inside a magnetic field to generate an electric current in a similar way to a dynamo. Electricity is generated in DC form, so it then has to go through an inverter to be transformed into AC for use by the grid, or most commonly the lighting and appliances in a given property.

A wind turbine's blade acts in a similar way to the wing of an aeroplane. As the wind blows, a low-pressure pocket of air forms on the blade's downwind underside, pulling it towards the pocket. This pulling, or sucking, makes the rotor spin through force known as 'lift' which is stronger than what is called the 'drag' effect (the force of the wind against the front of the blades). Working together, the lift and drag forces will spin the rotor in a constant direction to generate electricity. The blades, or rotor diameter, are also a useful key to guessing the kW output of a small wind turbine. Just by looking at a turbine it's possible to gauge, if only roughly, its probable output since the rotor diameter (also known

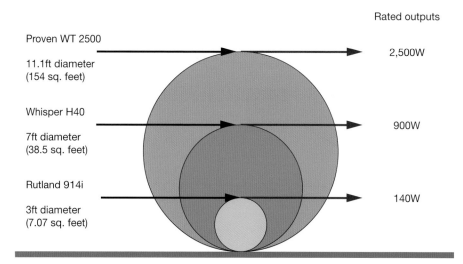

Figure 7.5 Swept area and near maximum power outputs of different sized wind turbines

Source: Dilwyn Jenkins

The wind power formula and cube law

Power = 0.5 p A v3 Cp watts
p = the density of the air (approximately 1.23 kilograms per cubic metre)
A = the area swept by the rotor or π r2
v = the speed of the wind in metres per second
Cp = the coefficient of performance
π = Pi (approximately 3.14)

The maximum theoretical amount of energy that can be extracted from the wind is 59 per cent – the Betz Limit. In practice 30 per cent is quite good.

A worked example:

A wind turbine has a rotor diameter of 2 m (6.5 ft) and the wind speed is 5 m/s. If the turbine's coefficient of performance at 5 m/s is 0.34, its output would be calculated like this:

Power = 0.5 × 1.23 × 3.14 × 5 × 5 × 5 × 0.34 = 82 W

Additionally, for wind power it is important to understand the cube law, which represents the relationship between wind speed and a given turbine's output. The example below shows us how significant the output increase is for every extra metre per second of wind speed:

Wind speed m/s	Turbine output watts	Rise in output	% Rise
5	82		
6	142	60	73
7	225	83	58
8	336	111	49

as the 'swept area' because the blades sweep it as they spin) more or less determines how much wind will be captured by a turbine.

Site Assessment and Wind Monitoring

Having as accurate as possible a figure for the wind resource at your turbine location is vital to deciding whether to invest in wind power, and which size and model turbine to select. Since access to the best possible wind resource (mainly a factor of location and hub height) is so important for determining potential output for a turbine, the first task is to decide where to monitor the wind. This should be at the best site close to the building and, ideally, a monitoring mast will be set up at the projected turbine hub height. With monitored data for one year or more it is possible to predict a turbine's annual output at the site with some accuracy. Most of the information below will help to establish where a potential wind site might exist and where it would be a good site to monitor the wind.

Finding the Best Wind and Wind Monitoring

The amount of wind required to power a turbine relates mainly to the turbine location – exposure of site, height of hub, local obstructions etc. A flag flying stiff in the wind is always a good sign. It is critical is to get the turbine hub as high as possible on the best possible site, not least because getting power from the wind has an exponential relationship with wind speed (e.g. a site 10 m above ground with 5.6 m/s wind speeds will generate twice the electricity as one with wind speeds of only 4.5 m/s). The US DOE say that to raise a 10 kW generator from an 18 m (60 ft) tower height to a 30 m (100 ft) only involves an extra 10 per cent in costs but could add 25 per cent to the power output. The importance of locating the turbine cannot be overstressed, both in terms of position in relation to topography and obstructions as well as the height of the tower or turbine hub.

Wind turbines don't like turbulent or gusty winds because they have to spend time hunting for which direction it is coming from every time the wind shifts. This puts additional stresses on the bearings, blades and yaw and means that the turbine will not always be spinning at the optimum speed to make best use of strong, but sudden, gusts.

Obstacles like buildings in the path of a wind actually divert it further and stronger than one might imagine. As oncoming wind hits a building a process called 'vortex shredding' happens, effectively dividing the wind into several

Table 7.1 Effects of the wind at different wind speeds

Wind speed			Wind effect
Meters per second (m/s)	Knots	Feet per second (f/s)	
0	0	0	Smoke rises vertically
0.3–0.6	1.9–3.8	1–2	Direction of wind shown by smoke drift, but not by wind vanes
0.6–1.2	3.8–7.8	2–4	Wind felt on face, leaves rustle and wind vane moved by wind
1.2–1.8	7.8–11.6	4–6	Leaves and twigs in constant motion and wind lifts a light flag
1.8–2.7	11.6–17.5	6–9	Raises dust, picks up papers and moves small branches
2.7–3.6	17.5–23.3	9–12	Small trees in leaf begin to sway and crested wavelets form on inland waters
3.6–4.6	23.3–29.1	12–15	Large branches in motion, whistling heard in power lines and umbrella use tricky

Source: adapted from the Beaufort Scale (devised by Admiral Beaufort of the British Navy in 1805)

different swirling flows that sometimes extend hundreds of metres beyond the original obstacle. The main local factors affecting wind speed, wind behaviour, turbine location and turbine selection are:

- topography;
- buildings;
- trees;
- prevailing wind direction;
- recorded average wind speeds;
- minimal gusting (turbines operate most effectively with constant wind and gusting winds will wear out the turbine parts much faster); and
- actual siting (as high as possible to catch the wind, especially given the exponential relationship between wind speed and energy output).

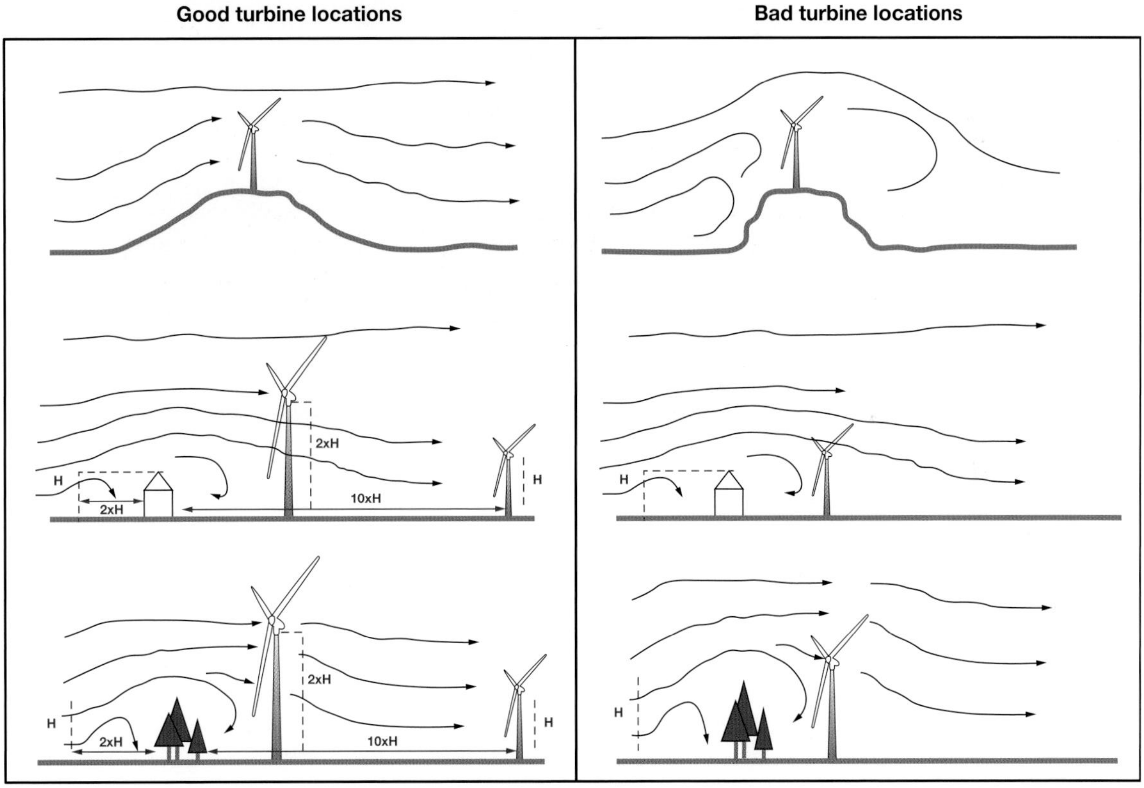

Figure 7.6 The zone of turbulence (i.e. the action of wind around local obstacles such as trees and buildings). Note that the zone of turbulence extends ten times behind the height of the obstacles (house, trees) – some sources give this as being 20 times the height of the obstacles.

Source: Dilwyn Jenkins (adapted from various previous examples)

Figure 7.7 This wind turbine is not ideally sited – it is frequently hunting the wind direction due to high turbulence because it is sited very close to buildings and right next to a cliff

Source: Dilwyn Jenkins

Measuring the Wind Resource

Maps and meteorological-office data can be useful: not to tell one where to install a wind turbine, but where there are good areas to monitor. Assessing the wind resource is much more complex than assessing the solar resource. Once one has solar data for a region that is usually enough, and as long as the PV array is installed at the correct angle and orientation and not in the shade the output can be predicted reasonably accurately. However, that is not true for a wind turbine: a machine installed a hundred yards away from another can perform very differently.

It is good practice to monitor wind speeds at the site where a wind turbine is to be installed for a period of 12 months although sometimes, in the case of smaller turbines, monitoring during the months of least wind might suffice. Data on average wind speeds is not enough. Reasonably accurate figures on the windspeed distribution throughout the year are required, and one needs to bear in mind that average wind speeds can vary by as much as 25 per cent from one year to the next. Other factors that affect the quantity of the energy that can be extracted from the wind are the density of the air (determined by humidity and altitude, the denser the air the more kinetic energy it contains), the contour of the landscape (sheer factor) and turbulence cause by vegetation and buildings (see Annex 2 for recommended books and further information). Perhaps the simplest option, if it exists, is to ask a nearby wind turbine owner for any local wind speed data they might have available. In an area where there's an airport nearby, it may be that by contacting their meteorological office they will be able to provide data that is of value.

A more local, but still simple, method of windspeed assessment is that of checking tree 'flagging' in the area immediately around the planned turbine site.

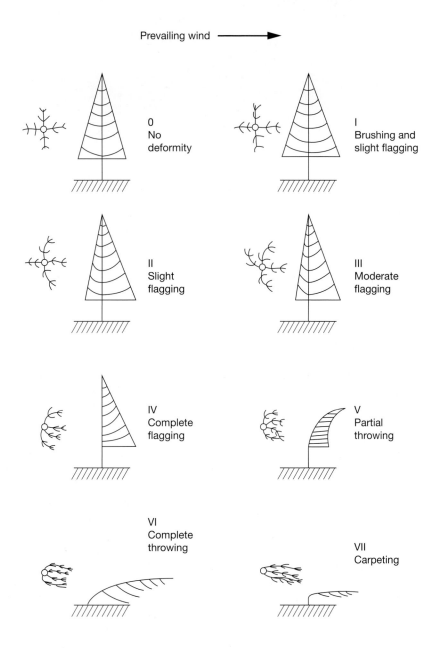

Figure 7.8
Windspeed rating scale based on the shape of tree crown and the degree to which twigs, branches and trunk are bent

Source: Griggs-Putnam Index of Deformity, after Hewson, Wade and Butler 1977

Figure 7.9 Photograph of flagging tree (somewhere between IV and V in the scale given in Figure 7.8 – complete flagging towards partial throwing)

A report on 'Wind Resource Assessment for Small Wind Energy Conversion Systems', published by the National Technical Information Service, outlines a number of wind resource assessment techniques, including 'flagging' interpretation under the Griggs-Putnum Index of Deformity. Flagging interpretation is based on the direct observation of vegetation, particularly trees (preferably evergreen). In windy spots trees are deformed by the action over time of the prevailing wind. These physical deformities have been categorised and correlated to specific windspeeds. There are seven flagging categories that are commonly used.

A doubling of wind speed results in an eightfold increase in available power. As a rough guide, for most turbines to pay their way you should have a minimum wind resource of 4 to 4.5 m/s (9 to 10 mph). Many would argue for a higher minimum wind speed of around 5 m/s to 6 m/s to ensure a reasonable payback scenario. However, average annual wind speeds can vary by up to 25 per cent. The important point is to have sound wind monitoring data and to match this to the power curve of a particular turbine.

Less vital, but also a factor (since different turbines start generating electricity at different cut-in wind speeds), is that wind speeds and cut-in speeds should be analysed together to work out the likely yearly output on any wind installation. In this book, wind speed is represented in metres per second (1 m/s = approximately 2.99 ft/s or 1.944 knots). In North America the wind is strongest (mainly between 7 m/s and 9 m/s) in the Rocky Mountains, particularly the eastern edge and down onto the Great Plains, where water pumping has been going on for well over 100 years. There are also strong wind areas just to the south of Chicago and to the northwest of Minneapolis (both areas achieving speeds of up to 7.5 m/s). This can be seen quite clearly on the Wind Powering America website (www.windpoweringamerica.gov/wind_maps.asp).

The average wind speed in the UK is around 5.6 m/s, little more than a gentle breeze. With this in mind, many small turbines are designed to cut in and generate from around 4 m/s. However, the best sites tend to be coastal or upland. The west coast of the UK has one of the best wind regimes in Europe.

The idea is to get the turbine hub as high as possible on the best possible site, not least because getting power from the wind has an exponential relationship with wind speed (e.g. a site 10 m above ground with 5.6 m/s wind speeds will generate twice the electricity as one with only 4.5 m/s). Locating the turbine is very important both in terms of its position in relation to topography and obstructions as well as the height of the tower or turbine hub.

Wind speed measurements should relate to the specific location and proposed turbine-hub height. This can only be done accurately with an anemometer, not by simply investigating generalised area weather data. Hand-held anemometers are available for less than US$50 and reasonably good anemometers can be bought for under $400, although the tower and associated data logging equipment could cost significantly more again.

Before investing in wind monitoring for a prospective turbine location it is worth checking available wind resource maps. Generally speaking, in the US the coasts, mountain ridges and the Great Plains have good wind speeds, but to obtain official wind data for every state there is the NASA World Wind data online, plus another online climate data directory run by the NOAA Satellite and Information Service (see Annex 2). There are also some useful government websites: one from the US DOE that provides wind assessments by US state called Wind Powering America (DOE Wind Resources), with quick links to state wind resource data; and the Wind Energy Resource Atlas of the United States (DOE).

In the UK, the Numerical Objective Analysis Boundary Layer (NOABL, a UK government initiative) can give rough average windspeed predictions. Wind speeds of 6 m/s should be identified before considering the installation of wind monitoring equipment. Most UK installers, however, recommend that a potential site have a minimum average NOABL wind speed prediction of 5 m/s. NOABL

Figure 7.10 Inside a wind-monitoring data logger

Source: Dulas Ltd

Figure 7.11 Wind-monitoring mast with solar powered data logger unit attached, which can be downloaded remotely via mobile phone connection

Source: Dulas Ltd

provides annual average wind speeds for a 1 km UK grid square. However, according to EST field trials (which looked at micro-scale domestic turbines, both free-standing and building-mounted, with a rated output of 400 to 600 W), the NOABL data has been shown to overestimate the potential wind speed at many sites, especially those in urban and suburban locations, partly because it is not sophisticated enough to consider the impact of local obstructions, including trees and buildings, in its methodology. The Carbon Trust has an online wind-yield estimator based on UK Met Office data. However, this is not sophisticated enough to account for local variations caused by structures or topography, and so is only a fairly rough guide.

Wind Turbine Technology

Once you have precise (monitored) data of the wind speeds for your site, the output for your selected turbine/s (how much of the kinetic potential of the wind can be converted into electricity) can be calculated using manufacturers' claims or actual performance data. However, it is worth noting that some manufacturers have been found, in a field trial by the UK's Energy Saving Trust, to overrate their turbines, so figures should be used with caution. In the trials, measured peak efficiencies of the turbines monitored in the field trial ranged from 30 to 45 per cent. Even large turbines, whose outputs are in the range of MW, only achieve peak efficiencies of around 40 per cent and the theoretical maximum efficiency of a wind turbine is 59 per cent (this is a physical constraint, called the Betz limit).

Definitions

These definitions are the same for all generation systems, not just renewable ones. Unfortunately, some reports are now incorporating the Capacity Factor into the Declared Net Capacity for wind farms, thus causing further confusion.

In this section, consider a commercial turbine rated at 1MW.

Efficiency: this is the ratio of actual output from a wind turbine compared to the energy in the wind. The theoretical maximum efficiency is around 59 per cent ('The Betz Limit') and commercial turbines achieve around 40 per cent. However, as wind energy is abundant, efficiency is not as relevant as when considering non-renewable energy sources. Our 1 MW turbine produces 1 MW in optimum conditions, and that figure accounts for its efficiency.

Declared net capacity (DNC): this is the maximum output you would expect from a connected turbine (or wind farm). It is 'net' because it allows for the energy used to run the turbine and any electrical conversion equipment on the site. If our turbine, when operating, uses 50 kW to run motors, lighting, controllers, etc. then its DNC will be 950 kW (i.e. 1 MW minus 50kW).

Load factor: this is the ratio of actual to peak output for a given period of time. For our turbine, if it is producing 425 kW at some point, then it has a load factor of 50 per cent (i.e. 425/950 × 100). If the period is a full year, this is the same as the *capacity factor* and the two terms are often used synonymously.

Capacity factor: this is the measure that takes account of varying wind speed over a year. Typically, in the UK the *capacity factor* is around 30 per cent. If our site matches the UK average, our turbine with a DNC of 950 kW will output, on average, 285 kW (i.e. 950 × 0.3). However, this is not the same as saying that the turbine will only operate for 30 per cent of the time. Typically, it will run more than 80 per cent of the time, but not always at its rated output. Note that using this average for your calculation is no substitute for measuring the wind speed at your site over a long period and doing a full yield analysis.

As wind energy is abundant, and free, and for wind power the rated capacity is just a nominal output at a chosen windspeed (usually the maximum output under normal conditions: i.e. the peak of the power curve), the concepts of efficiency and load factors don't mean that energy is being 'wasted' or that the turbine is underperforming. They are only relevant to giving an accurate picture of how the turbine can be expected to generate.

With variable-speed wind turbines the rotor spins a brushless AC generator to produce AC electricity (single- or three-phase). The voltage and speed of the AC electricity produced depends on the spinning speed of the rotor which, in turn, depends on the speed of the wind. This is not very useful either for feeding into the grid or for charging batteries, so this variable AC is rectified to DC. Most small turbines contain these rectifiers in the nacelle. Larger machines convert from AC to DC in the controller located away from the turbine.

Power Curves

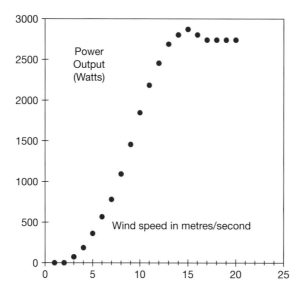

Figure 7.12 Power curve for Proven WT2500 wind generator

Source: Proven Energy data, www.provenenergy.co.uk

Power curves for wind generators are essentially graphs indicating how much power the turbine is calculated to generate within a range of given wind speeds. This power is the turbine's output in W or kW and is shown on the vertical axis, while the windspeed range is expressed along the horizontal axis. Most turbine manufacturers have their own power curves for each model of turbine that they produce. According to the results of the UK's Energy Saving Trust's (EST) field trials, mainly with micro wind generators, the manufacturers' power curves are not always particularly accurate, with the tendency being towards over-optimism. With this in mind, it's sensible to introduce a margin of pessimism (e.g. 10 per cent less power-output than suggested in the manufacturer's power curve at a given windspeed) into any calculations for paybacks based on a particular wind turbine at an estimated average annual windspeed. More independent testing is needed, but to do this requires live windspeed monitoring as well as data logging of the turbines' outputs.

Micro, Small and Medium-sized Machines

In this section we explore the range and main types of wind turbines available on the market and potentially appropriate for use in the home or business situation. As described above, the very smallest wind turbines – micro wind generators – tend to be designed for the leisure market: boats, caravans and camping. They provide enough power to trickle charge a 12 V battery, or perhaps for a small lighting system. Small-scale generators, on the other hand, are generally sized to provide a significant part (or all) of the electrical needs of a household or business unit. Medium-sized wind turbines – between 50 kW and 500 kW – can generate significant amounts of electricity and are more usually associated with community wind, factory units or a straightforward commercial venture selling electricity direct to the grid.

The definition of small wind turbines actually covers quite a scale range. Anything from a 50 W battery-charging wind generator up to a 10 kW free-standing turbine would fit this category. It's important to bear in mind that the kW rating of a turbine is generally what it will produce in optimum conditions (e.g. 11 m/s winds). Even the most windy locations are highly unlikely to operate in such optimum conditions very often. In the sections below, we take a closer look at their respective uses and track records.

USA Consumer Guide recommendations and UK fields trials (EST, 2007)

(http://www.energysavingtrust. org.uk/Generate-your-own-energy/Energy-Saving-Trust-field-trial-of-domestic-wind-turbines)

The USA small wind turbine Consumer Guide recommends the following key circumstances to consider even before monitoring your wind:

- Is there a good enough wind resource?
- Are tall towers acceptable or permitted in your locality?
- How much electricity do you need (either for own use and/or to sell to the grid)?
- Can you connect to the grid in order to sell it on (or is the grid so far away that wind power is a stand-alone alternative)?
- What will the costs of installing and maintaining the turbine system be?

One other point could well be added:

- What are the locally available levels of capital grant incentives and preferential tariff rates for the sale of clean electricity and tax incentives?

Recommendations of the EST Field Study include these guidelines for potential turbine buyers:

- Wind turbines work only when installed properly in an appropriate location.
- They offer potential for delivering carbon savings and energy generation.
- The highest potential for successful household small-scale wind installations in the UK is in Scotland (because of the strong and consistent winds).
- Wind speeds are difficult to predict and highly variable.
- Utilise the best available windspeed estimation tools and, if possible, use anemometry to determine the windspeed distribution (for more on this see Basic Physics and Wind Monitoring sections above).
- Buyers are advised to only consider domestic small-scale wind products and installers that are certified under the Microgeneration Certification Scheme (applies to UK only).
- Householders should consider energy produced from small-scale wind as one option from a potential suite of microgeneration technologies.

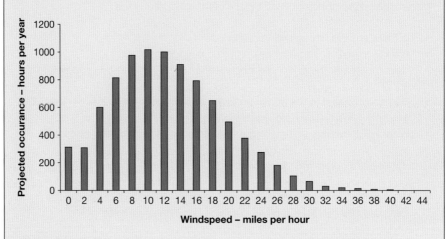

Figure 7.13 Example windspeed distribution at 13 mph average windspeed (1 mph = 0.44704m/s)

Micro wind generators (under 1 kW)

Whilst micro generation is often defined as anything under 50 kW, and some wind turbines in the 1 kW to 5 kW range are sometimes described as micro generators, in this book we only consider the smallest turbines as micro wind generators. They are primarily used to provide remote power for leisure activities such as boating or caravanning, or for other locations where independence from the grid is important but power requirements are relatively low. They are ideal for bringing the benefits of electricity to a wide range of remote applications: camps, garden sheds, remote monitoring and laptops etc. Some are also integrated or tied to building structures, a highly controversial issue that we will look at later in this chapter (attempts to bolt even very small-scale wind generators to

the side or roof of the average house or conventional larger building does not have a very good track record).

In the US, companies such as Southwest Windpower and Bergey have been in operation since the late 1970s and 1980s. Southwest Windpower started off in 1987 making small battery-charging wind turbines to complement solar energy systems in rural areas. Within seven years the bestselling Air battery-charging turbine had been installed in over 100 countries. The company currently makes Sky, Air and Whisper turbines for marine, residential and commercial applications. Bergey Windpower's smallest turbine, the XL1, is also designed for off-grid battery-charging applications and boasts quiet operation achieved by an oversized alternator that reduces the rotor operating speed

The UK company Marlec, active in renewable energy since 1979, has a long and successful track record in the field of micro wind generator manufacture and sales. Their classic model – the Rutland 913 – is renowned for being quiet and is designed for optimum performance from the lowest of wind speeds to surviving gales. These are typically seen on yachts, but are also used to power remote functions such as street-lights, road signs and off-grid CCTV. In the same range, the Rutland 914i offers 30 per cent greater outputs. They also make a 910 (or Furlamatic) that automatically turns the turbine away from what might be otherwise devastatingly strong gusts and winds. Marlec also offer grid-connect systems, some often using building integrated PV.

Since the 1970s, another British company – Ampair – have produced a variety of custom wind power supplies, many for deployment in some quite inhospitable climates. With more than 40 years in the industry and over 30,000 systems already sold, like the Marlec Rutland windchargers they have proven reliability. Ampair wind generators range from under 100 W peak through 600 W and up to 6 kW, the latter aimed at grid-connect situations.

Figure 7.14a The Air 40 is marketed as a small battery-charging turbine for remote homes, telecom, industry, lighting etc.

Source: Southwest Windpower, www.windenergy.com

Figure 7.14b An installation of three turbines in Beijing, China (these are battery-charging wind turbines produced by US company Bergey)

Source: Bergey Windpower, www.bergey.com

Figure 7.15 Rutland Windcharger (Marlec) being used in conjunction with a solar photovoltaic panel to power street lighting – the low friction three-phase alternator gives a smooth and silent output and the low windspeed start up means that even light breezes will provide some generation

Source: Marlec, www.marlec.co.uk

Small wind turbines (1 kW to 15 kW)

Wind generators with rated outputs of over 1 kW, but less than 15 kW, fall beyond the micro range and within the category of small turbines. These are likely to have rotor diameters of some metres and usually require towers of between 15 m (50 ft) and 30 m (100 ft). Apart from a few options (including some vertical axis machines designed for urban situations), this type of turbine is the most appropriate for powering homes or businesses, and is most commonly used in combination with free-standing towers. They are suitable for both grid-connect and stand alone systems. To date, most turbines in this range will be found in windy, probably remote, areas, but the increasing opportunities offered by preferential feed-in tariffs, net metering, two-way metering and utility buy-back schemes is beginning to make these small turbines financially as well as environmentally advantageous even when the grid is locally available.

However, the issue of finding a windy enough and otherwise appropriate site for the turbine is a major factor to consider, more so than with micro turbines. Even if the site has promising wind speeds, the problems of locating the turbine where it will suffer minimum turbulence from local wind obstructions (houses, trees, steep cliffs etc.) and not cause issues for the neighbours can be more serious than with a micro wind generator.

Having expressed that point, it is fair to say that even at the smaller end of this range (1 kW to 10 kW), small wind turbines can make significant contributions to the electricity consumed in a home or business. At the upper end of the range there may well be substantial surplus electricity produced which can be exported and sold to the grid. According to the US DOE, a 1.5 kW wind turbine at a location with average wind speeds of 6.26 m/s could provide around 300 kWh a month, sufficient for most electricity-efficient homes. The same source suggests that a typical US home uses approximately 9400 kWh of electricity per year (about 780 kWh per month). This means that, depending on the average windspeed in the area, a wind turbine rated in the range of 5 to 15 kW would be required to make a significant contribution to this demand, particularly if combined with energy efficiency measures.

There are over 300 Evance Iskra R9000 5.3 kW wind turbines (formerly called the Iskra turbines) in the field, in a range of situations including rural domestic properties, schools, small farms and light industrial sites. The manufacturers supply the turbine with free-standing 'monopole' or guyed supports that require adequate space. This is ideal for power entry into the grid, but the 400 V output can be inverted to 240 V for domestic use. The Evance R9000 has a blade diameter of 5.4 m. This means it has a swept area, or capture area, of 22.9 m². This turbine has been used in community wind schemes.

Figure 7.16a Bergey's 10 kW Excel turbine, installed near the Capitol Building in Oklahoma City, US

Source: Bergey Windpower www.bergey.com

Figure 7.16b The two-bladed Whisper 500 from Southwest Windpower. Its fibre-glass re-enforced design is designed for harsh environments and high windspeeds

Source: Southwest Windpower, www.windenergy.com

Table 7.2 Comparing two small Proven wind turbines: the Proven 11 (5 kW) and the Proven 35–2 (12 kW)

Proven turbine	11	35–2
Rated power (at 11m/s)	5.2 kW	21.1 kW
Peak power	6.1 kW	13.7 kW
Rotor diameter	5.5 m	8.5 m
Application	Consumers of up to 9,000 kWh a year (e.g. commercial buildings, farms and government offices)	Consumers of up to 23,000 kWh a year (e.g. commercial and government buildings, large farms and for sale to grid)

Note: At the time of writing, Proven were owned by Kingspan and were selling the KW3 and KW6 (http://www.wind.kingspan.com).

Source: adapted from Proven Energy data, www.provenenergy.co.uk

Figure 7.17a, 7.17b and 7.17c Evance Iskra R9000 5.3 kW (formerly called the Iskra) turbines are used in a range of situations, including rural domestic properties, schools, small farms and light industrial sites

Source: (a) Hockerton Housing Project, www.hockertonhousingproject.org.uk, (b) and (c) Dulas Ltd, www.dulas.org.uk

WIND ENERGY 153

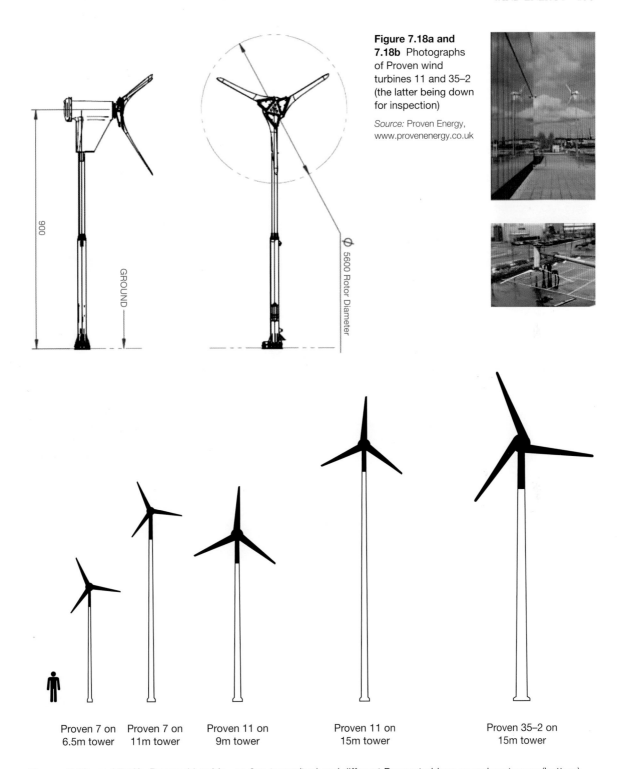

Figure 7.18a and 7.18b Photographs of Proven wind turbines 11 and 35–2 (the latter being down for inspection)

Source: Proven Energy, www.provenenergy.co.uk

Proven 7 on 6.5m tower   Proven 7 on 11m tower   Proven 11 on 9m tower   Proven 11 on 15m tower   Proven 35–2 on 15m tower

Figures 7.19a and 7.19b Proven 11 turbine on 9 m tower (top) and different Proven turbines on various towers (bottom)

Source: Proven Energy, www.provenenergy.co.uk

Medium-scale wind turbines (15 kW to 500 kW)

Medium-scale turbines range in output from around 15 kW to about 500 kW. They tend to have rotor diameters in the 20 m to 30 m range and towers of between 30 m and 70 m. These are ideal for many community-based wind power schemes, both for off-setting the electricity of a group of houses or business units and for selling surplus to the grid. Over about 15 kW to 20 kW, most wind turbines are likely to be developed primarily in order to generate income from the sale of electricity to the grid.

Figure 7.20 Endurance E-3120 50 KW – a three-bladed horizontal axis downwind turbine – is ideal for larger farms, schools, hospitals and commercial/industrial sites

Source: Dulas Ltd, www.dulas.org.uk

Figure 7.21 Turbowinds T500–48 500 kW wind turbine, ideal for business parks and grid sales

Source: Turbowinds, www.turbowinds.net

Endurance E-3120 50 KW will produce 100,000 to 250,000 kWh per year in appropriate winds. Designed for direct grid-tie, the Endurance 50 kW machine has a cut-in wind speed of 3.5 m/s and a rotor diameter of 19.2 m (giving a swept area of 290m^2). It has a programmable logic controller and either a wireless or wired network software interface for remote monitoring.

The Turbowinds T500–48, 500 kW wind turbine is one example of the higher end of medium-scale turbines. It produces enough energy to power up to around 100 homes, but this clearly depends on how much energy the households use (e.g. homes in the US tend to use up to twice as much electricity as those in the UK, and many businesses use less than the average home). These are not designed as residential turbines. Community-sized, they produce the right amount of power for school and university campuses, residential developments, farms, municipalities and business parks. They can also be used in small wind farms just for grid electricity sales. They also meet the requirements for the UK's feed-in tariff.

Larger turbines (500 kW and bigger)

Although not the subject of this book, larger wind turbines (those that are 500 kW or bigger) are where most of the R&D investment has been made in wind power technology over the last 25 years or so. There is about 200 GW of wind power now installed across the globe, mostly in clusters of turbines or wind farms. The visual impact of large wind farms has slowed the uptake of this technology through the planning process, particularly in the UK which is a relatively small and densely populated island. Wind farms, however, can (and in many places already do) make a significant contribution to the overall supply of electricity,

Figure 7.22 Large-scale wind turbines don't seem to bother livestock and function well in windy and rural areas

Source: Dulas Ltd, www.dulas.org.uk

Figure 7.23 Large-scale turbines need to be located in windy areas, often found in upland or coastal sites

Source: Dulas Ltd., www.dulas.org.uk

Figure 7.24 Many of the newest wind farms are located offshore – this creates a new set of engineering issues for developers and even monitoring the wind can be tricky

Source: Dulas Ltd, www.dulas.org.uk

and while the wind may not always blow strongly, this natural power-source uses the grid as a store. Electricity from wind farms can be moved around from remote windy areas to urban and industrial zones easily via national grid networks.

Economics

The economics of a wind turbine depend on three main factors: investment costs, turbine output and any incentives that are available, either for capital costs or preferential tariffs for the sale of electricity to the grid. As discussed above, the turbine output depends on the wind regime at the rotor hub and the specifications of the turbine selected. Towers and cable runs form part of the overall costs, as does wind monitoring to prove feasibility of the location and any inverter, control equipment or grid-linking devices.

Towers

Towers are almost always supplied by the turbine supplier. They are a major cost item, sometimes accounting for a greater proportion of the overall wind-system investment than the turbine itself. Towers are also quite heavy. A 9 m tower for the Proven 11 turbine weighs 678 kg (78 kg more than the turbine). A 15 m tower for the same turbine weighs almost twice as much at 1200 kg.

There are two main types of tower: free-standing and guyed. The free-standing towers are usually more expensive, larger and self-supporting units, embedded into concrete foundations. Guyed towers are supported by cables attached to the ground and the tower and are generally cheaper and more varied in design, ranging from single pole (pipe or tubing) installations to simple lattice towers. Typically, for strength and safety factors, a guy's radius is designed to be between 50 and 75 per cent of the tower height. These are simpler to install than free-standing towers. Some of these have a tilt-down facility making them even easier to erect and faster to bring down (for maintenance or before a big storm), but only really useful for smaller turbines (5 kW or below). Towers range in height from under 10 m to over 100 m (the latter for large-scale commercial turbines that might generate as much as 2 MWp).

The simplest method for micro wind generators and small turbines on light-weight tubular towers is with a gin pole. Essentially a lever, it is erected at the base of the turbine with a lifting wire attached to the top of the turbine. Guys are attached to the top of the pole, and at the turbine anchor points, before the gin pole is lowered by pulling the lifting wire (usually using a vehicle). Guys can be detached and used to anchor the turbine and tower while the lifting wires are removed. Raising tubular towers, and even some lattice towers, can be achieved via hinges at the base which also help post-installation by facilitating any dropping and re-lifting for installation and maintenance or repairs. Aluminium towers are not really strong enough and are best avoided. Really large turbines with massive free-standing steel towers are generally lifted onto their concrete foundations using a crane and gantry, often in cylindrical segments.

Turbine Siting and Cable Runs

Due to the factors already discussed – including local topography, trees and man-made obstructions to the wind – in most circumstances a small wind turbine is ideally located at least 10 m above any surrounding obstacles within a 100 m radius. However, this isn't always possible. The cable run between turbine and load also needs to be considered, since there will be some power loss and it is another cost factor.

Incentives

A system of 'net metering' is operated by many US utilities to allow clean energy generators, even small ones, to use electricity meters that can effectively run backwards when the property is generating more electricity from the wind than it is consuming at a given time. This allows user-generators to offset consumption throughout the entire billing period. Each state and utility has a slightly different way of operating net-metering schemes, some with a system of credits that allow a small generator to spread its costs and income through the year. For any building that uses winter wind to displace large summer cooling loads, this can help smooth out the cash-flow between electricity purchase costs and income from generation.

In the UK, wind power incentives mainly take the form of preferential rates for clean energy, feed-in tariffs (see relevant section below) and the Renewable Heat Incentive (RHI, see Chapter 3). Presently, UK domestic customers with a wind turbine can sell the excess electricity generated back into the grid through an arrangement with the energy supplier and regional electricity Distribution Network Operator (DNO). 'Buy-back' rates vary between different energy suppliers, so it's difficult to make accurate financial predictions about income flows from the sale of electricity without first getting an offer from your utility company.

Turbine Costs and Buying a Turbine

The Canadian Wind Energy Association suggests that domestic wind turbines (up to 10 kW in size) cost between CAN$3,000 and $4,000 per kW, while the total installed cost (including works, tower, inverter, controls, cables etc.) is double this. For larger applications, suitable for farms and businesses (10 kW to 300 kW), they estimate turbine-only costs to be between $2,000 and $2,500, with the total cost at between $3,000 and $4,000.

Average 1 kW turbine costs in the UK range from around £1,500, but with the full costs of installation the final figure is likely to be closer to £3,000. Systems up to 5 kW or 6 kW generally cost up to about £16,000. Larger turbines, as a rule, cost less per kW output than smaller ones since there are economies of scale for elements such as cable runs, tower installation, professional services, inverters and grid-connection costs, etc. The cost of the turbines themselves are likely to come down over the next decade or two as pressure mounts to mitigate and resolve global climate change by encouraging more R&D and greater scales of production in wind and other renewable energy sources.

Table 7.3 A rough guide to wind-turbine system costs

Wind system cost items	Approx. cost (%)	Comment
Wind turbine generator	40%	Erected, commissioned and with warranties, guarantees and system manual
Tower and foundations	40%	Erected, commissioned and with warranties and guarantees
Inverter and grid tie equipment	5%	Installed and commissioned
Cables (includes laying)	5%	Might involve posts and/or trench digging
Pre-decision wind monitoring	3%	Either professionally executed or DIY
Transport & delivery	6%	Particularly relevant if a large tower structure is required
Permissions	1%	Varies with region and scale

Source: adapted from various sources by Dilwyn Jenkins

These days many turbine installations, particularly ones that are for domestic or small business use, are likely to qualify for government grants, often up to a third or even half the capital costs being covered. This can effectively half the payback period. A well sited 6 kW turbine can be expected to generate around 10,000 kWh per year, offsetting over 5 tons of CO_2 annually and, if part of the UK's feed-in tariff scheme, even generating sales to the grid of over £3,000 a year. For properties with turbines not selling to the grid, there are still substantial potential savings on electricity (at 12p a unit for 10,000 kW = £1,200 a year).

In terms of calculating how fast your payback might be, it's worth remembering that peak power for small turbines is generally greater than its rated power; a Bergey XL1 is rated at 1000 W with a wind speed of 11 m/s (24.6 mph) even though it is technically capable of generating outputs of 1300 W at higher windspeeds. If your wind speed is only 7 or 8 m/s, the output will be significantly lower. A much more accurate assessment can be made simply by comparing the wind monitoring data for the site at rotor-hub height with the selected turbine's output power curve.

Apart from the costs of buying and installing a turbine and tower – along with the necessary cabling, control systems and inverters – wind devices need regular maintenance checks, some more than once a year, others every ten years or so. If well maintained, a good wind turbine can be expected to last 20 to 25 years. Blades and bearings might need replacing every 8 to 12 years and need to be checked annually. However, off-grid wind systems need a bank of batteries to store the energy for using when it's needed. Battery life is typically between five and ten years.

The first steps to consider before actually buying a turbine:

- Have you reduced your property's electricity demands through energy conservation and energy efficiency measures?
- Find out what the average annual wind speed for your site might be.
- Assess the local topography for obstructions and advantageous potential turbine locations.
- Calculate what size turbine might be appropriate for your property.
- Check with your local authority as to whether planning or building-control approval are necessary (including for any wind monitoring tower and equipment), and, if they are, whether it is likely to be given.
- Think about the neighbours: noise, vibrations and some visual impact are often caused by wind turbines of all sizes.
- Monitor wind potential at the site.
- Investigate potential metering arrangements with the utility.
- Research and contact selected manufacturers and/or supplier–installers. Both should be certified or approved under the relevant scheme (e.g. the US net metering programme or, in the UK, the Microgeneration Certification Scheme).
- Check out grant eligibility and incentives such as payback or feed-in tariffs).
- Ask for and assess different suppliers' terms and conditions (including warranties, guarantees and the proposed schedule of deposits and payments).
- Analyse how much the selected model/s would cost to install and maintain (including what grants are available) and how much electricity is likely to be produced at your site.

The American Wind Energy Association (AWEA) provides an online list of equipment manufacturers and dealers plus a list of recommended questions that potential turbine buyers should remember to ask:

- What is the energy output (measured in kWh not in kW) of the turbine, over one year, in average wind speeds of 8 to 14 mph? Is this calculated using real-life ('field') data (preferred), or laboratory/wind tunnel testing?
- Can you refer me to other customers who have owned [Model X] for a period of time? (The longer, the better.)
- What is the warranty length and coverage (industry standard is 5 years)?
- Has the turbine/tower ever gone through a reliability test? By whom? For how long? What were the results?
- How long has the company been producing turbines?
- How long has [Model X] turbine been available to sale to ordinary customers (i.e. not in the prototype or beta testing phase)?
- For how long was the prototype tested? By whom? In the field or in a laboratory?
- How many turbines of [Model X] been sold, and for how many years? How many of these are still running?
- How frequently has [Model X] been redesigned? What were those changes and how recent were they?
- What problems have other customers encountered and how have you dealt with them?

Connecting to the Grid and Feed-in Tariffs

If a wind system is designed to generate a surplus of electricity over and above the requirements of a home or business then this can, in most circumstances these days, be exported and sold into the national grid. A qualified or accredited installer/electrician is needed to connect a wind turbine's output to the mains in order to get a grant under most incentive schemes. For safety's sake, grid-

Table 7.4 Logical phases and steps in the process from feasibility assessment to installation of a wind turbine

Phase	Steps
1. Feasibility phase	Wind resource assessment
	Turbine location
	Research permissions and grid linking
	Budget estimations
2. & 3. System design & planning phases	Discuss with installer or professional
	Select turbine, tower, inverter and other system cost items
	Finalise financial calculations (including potential income)
4. Installation phase	Select competent installer
	Obtain utility connection and sale approval as well as permissions
	Competent team install and commission system

Source: Dilwyn Jenkins

connected wind systems will automatically shut down if there's a grid power-cut in the local electricity distribution network. The reason for this is that small wind turbines need batteries or a grid to smooth out fluctuations in the consumption and generation of electrical power.

In order to sell electricity into the grid you will need to invest in a special meter (about US$185 or £120) as well as negotiate a rate with your electricity supplier. Rates vary from supplier to supplier and also between state (in the US) and national (in Europe) policies, whether government or utility based. If the supply is liberalised and competitive, it should be possible to shop around and find a sympathetic supplier.

In the US there are sometimes small electricity buy-back credits available for wind power fed back into the grid. The rate and its availability varies from state to state and year to year, but information is available from your local utility. There are two programs relevant to small-scale power producers: the National Energy Policy Act (1992) and the Public Utilities Regulatory Policy Act (PURPA).

Australia operates a form of net metering, offering the standard feed-in tariff (FiT) for people producing power for their homes or small businesses using renewable energy systems with a capacity of up to 100 kW. This includes wind, solar, hydro or biomass power. Excess power fed back into the grid is credited at the same retail rate charged for electricity consumed. Australia also operates a Solar Credits mechanism under the expanded Renewable Energy Target (RET) that multiplies the number of Renewable Energy Credits (RECs) able to be created for eligible installations of Small Generating Units. New wind turbines are eligible under this scheme if under 10 kW, with a total annual electricity output less than 25 MWh. Solar Credits are linked to the RET system, where each REC represents 1 MWh of renewable generation and has a tradable dollar value. The Australian Renewable Energy Regulator publishes a 'Small Generators Owners Guide', which describes how to calculate how many RECs your system could earn.

FiTs have been available in the UK since 1 April 2010, but are not presently available in Northern Ireland. The scheme was designed to encourage domestic, business and community energy users to generate their own clean electricity. Several UK electricity supply companies – including the 'big six' – offer to buy wind-generated electricity. This scheme guarantees both a minimum payment for all electricity generated and a separate payment for the electricity exported to the grid. On top of this, the property will experience smaller bills, due to grid-imported electricity being offset by the wind generator's output. If the turbine is specified to produce more than the property consumes, over the year the bills should be positively in the favour of the generator. At the heart of the FiT is a special two-way metering concept that allows generators to sell to the grid at a higher rate per kWh than they buy from it. At the present UK rate, over the lifetime of a tariff a wind generator can expect to recover the capital investment more than twice over. FiTs are planned for a 20-year period for wind power. Since April 2010, the FiTS scheme has insisted on the use of a Microgeneration Certification Scheme (MCS) certificated product and installer (see Annexes 1 and 2 for more information on this).

Planning Permission and Regulations

If you are considering investing in a wind turbine installation, and well before investing in any wind monitoring at the site, one of the first essential tasks is to contact your state government, local authority or housing body regarding planning permission and the relevant regulations surrounding the installation.

In the US – particularly in a suburban environment – it's important to check the relevant local building codes and housing covenants. Some jurisdictions restrict the height of any structure in residential areas. Wherever you are considering siting a turbine, if possible talk to the neighbours first. Most restrictions in the US are going to be imposed at state or housing (urbanisation) organisation level in the form of a zoning ordinance, most of which have height limits of 35 ft (just under 11 m). A call to your local building inspector, boards of supervisors or planning board should resolve the issue one way or another.

Planning permission is still a genuine issue in the UK for micro and small wind generators despite the introduction in 2008 and 2009 of permitted development rights for renewable technologies. The permitted development rights basically lifted the requirements for planning permission for most domestic micro generation technologies. However, because of legal technicalities it does not cover micro or small wind. It is expected that these issues will soon be resolved and that both roof-mounted and free-standing wind turbines will be permitted at detached properties, although probably not in conservation areas. Further legislation is expected, but it will still be vital to consult your local authority about the need for planning permission and regarding building control.

In Canada, wind energy installations must comply with local building codes, applicable zoning laws and permit requirements. Check with local municipal offices for information about requirements and restrictions, including, where relevant: minimum lot or property size, required setbacks from property lines or neighbouring structures, height restrictions and noise ordinances. Also, you will need to demonstrate that your project complies with the applicable building and electric codes. The latter requires a line drawing showing conformance with existing electrical codes and the applicable authority (e.g. Electrical Safety Authority in Ontario). Small wind turbines require the notification of and approval from Transport Canada, local permitting agencies, and (in the case of grid-connected turbines) your local utility. A building permit, if also required, can take up to six weeks or more.

Transport Canada requires that prospective owners accurately provide the location (latitude and longitude) and height of all wind turbines to be installed so that they can be plotted for aeronautical maps and flight paths. Navigation Canada may then impose specific markings and lighting that must appear on the turbines so that helicopters and aeroplanes in both commercial pathways and search-and-rescue zones can visually identify them.

The Clean Energy Council is the Australian organisation that administers the accreditation of renewable energy installers and grid-connected inverters in Australia. Before buying a turbine in Australia it would be a good idea to enquire with your supplier about whether the turbine you are considering meets the interim Australian standard AS 61400.2(Int)-2006 wind turbines. In terms of planning regulations, the first port of call should be the State Government's Planning Department.

8
Hydropower

Introduction

While the term 'hydropower' evokes echoes of the past, it is still considered the renewable energy of choice if the right resources are available. The civil works and capital investment required for hydropower often look daunting in comparison to those of solar or wind power, but the benefits of hydro are clear. Firstly, a well-built hydro installation may last between 50 and 100 years, with little major maintenance required for at least the first 20 years. Taking the lifespan of comparable energy generating devices into account, hydro works out much cheaper per delivered kWh than wind or PV. Secondly, unlike solar and wind, hydropower can be available 24 hours a days, seven days a week, month after month, year after year. Unexpected droughts and changing climates aside, hydropower has the most continuous and predictable output of any renewable energy. On-grid it can continuously offset in-building electricity needs while selling surplus to the grid. Off-grid it can offer a continuous electricity supply. The old reputation of hydro schemes as unreliable and needing constant manual maintenance has been cast aside in the last 20 to 30 years, initially with the introduction of automatic electronic load-controllers and more recently with the proliferation of self-cleaning intake screens. If you have access to an appropriate water resource, hydro could well be the best of all renewable energy solutions.

In North America and Europe, preferential feed-in tariffs (FiTs) are increasingly available for clean electricity generation (see Annex 1 for more details). In combination with these tariffs, hydropower often becomes the most bankable of renewables. In some US states there are also rebates available which offset some of the capital costs. Hydropower will only work, however, if an appropriate water resource is available. Even with a good stream or river resource with plenty of drop, most countries have a series of often very tight regulations – implemented by National Water Authorities, Environment Agencies or Ministries – which can restrict a hydro project's generating potential and increase the capital costs to such an extent that it becomes non-viable.

Hydro presently provides more than 20 per cent of the world's electricity requirements and, in 2010, accounted for 6 per cent of total US electricity generation and some 60 per cent of renewable generation. Given the source of hydropower – water – it's not hard to understand why the term can encompass everything from mechanical water mills to ocean-based wave machines for generating grid electricity. In this chapter, however, we focus on the more established and traditional form of hydropower, which offers the potential to take energy out of streams and rivers as long as there is sufficient flow and drop (or 'head', see the Useful Hydro Terms section below). In particular, we focus on micro hydropower that can provide electricity in the range of 500 W to 5 kW. The same principles apply to larger micro and small hydro systems: it's just all on a larger scale.

> ## Hydro Scales (as used in this book)
>
> The definitions of terms and scales for hydro vary considerably country by country, but generally speaking micro hydro is understood as the application of hydroelectric power sized for small communities, single families or small enterprise. The term 'pico hydro' is also commonly used to describe hydro schemes in the range up to about 5 kW, which relates to most of the schemes or designs referred to in this chapter.
>
> **Micro hydro**: 0.5 kW to 100 kW (for use in local buildings and possible sale on to grid)
>
> **Small hydro**: 100 kW to 5 MW (to power communities or industrial process, usually via sale to grid)
>
> **Large hydro**: 5 MW to 50 MW (usually for national grid)
>
> **Mega hydro**: Greater than 50 MW (national grid)

The Hydrological Cycle

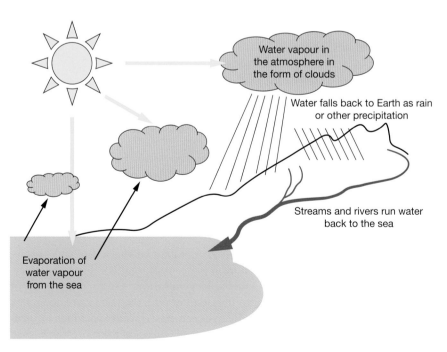

Figure 8.1 The solar-driven hydrological cycle

Source: Dilwyn Jenkins

Like wind power, bioenergy and most ground-sourced energy, hydro is essentially yet another form of solar power since the sun is the main driving force behind the hydrological cycle. The word *hydro* comes from *hydra*, Greek for water. For the ancient Greeks, Hydra was a terrifying, many-headed water beast that, like the white-headed waves of a tempestuous Mediterranean, was feared and demanded respect. However, the hydrological cycle is one of the most basic life-bringing processes on which life on Earth depends. Quite simply, the sun's heat evaporates water in the form of vapour from the planet's water surfaces, mainly the oceans. Via winds, which are also driven by solar energy, the airborne water vapour is shifted around in the atmosphere, visible as clouds. In the right conditions – for example further condensing due to air-cooling or just bumping into a mountain range – the water vapour is dispersing back onto the Earth's surface as rain or other forms of precipitation.

Ever persistent in its intent to find a way back to the oceans, water molecules use gravity, topography and the relative softness of different soils and rocks to find their way. Any precipitation not directly absorbed by the biosphere (i.e. the plant, insect and animal layer of life) or subsurface of the Earth eventually merges in the world's millions of springs and streams. It then blends together into the tributaries and main arterial rivers that drain every valley and every continent. The force of this movement of water back to the oceans is the key to traditional hydropower.

Useful Hydro Terms

Head and *flow* are the two main ingredients required for determining the hydro energy that can potentially be captured from a stream or river. *Head* is defined as the vertical height that a given water-resource could theoretically fall through the turbine or wheel: usually the vertical difference between a hydro scheme's intake pipe opening (often at the level of a reservoir) and the location of the hydropower device. *Flow* is best defined as the potentially available water resource, or quantity of water available, which can be measured over units of time (e.g. metres per second or m/s).

A lake or reservoir of water is *potential hydro energy*. As water moves downhill, perhaps from the reservoir, it becomes *kinetic energy*, some of which is lost in its transformation into sound or friction as it weaves its way through the local terrain.

A Pocket History of Water Power

The ancient Greeks had already mythologised the obvious power of water as a common yet wild element, whose domestication by humans began more than 2,000 years ago. The ancient Egyptians used a simple horizontal axis water-power device called a *noria* wheel that was fitted with jugs or buckets to lift water from rivers and streams to higher surrounding land in order to irrigate crops. This simple mechanical form of hydropower was soon followed by what was at the time an innovative vertical axis water wheel, known as *ghata* wheels (still commonly used in Nepal) for grain milling. These were both *undershot wheels* where the blades are turned by water flowing at or near the base of the wheel, pushing it from below.

Figure 8.2 Making a crude (gross) 'head' measurement

Source: Dilwyn Jenkins

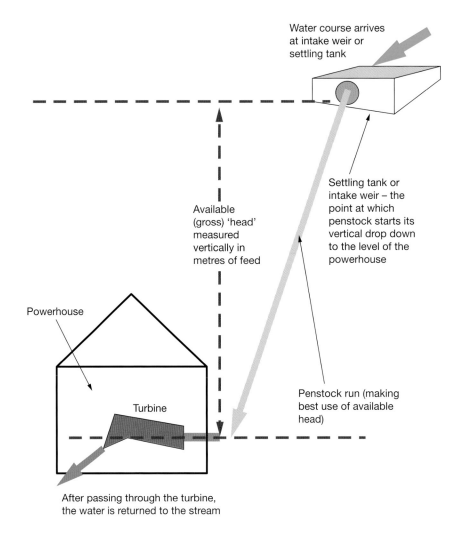

Between the eighteenth and early twentieth centuries, water power assisted the development of the industrial revolution and some initial provision of domestic and commercial electricity, particularly in rural areas of Europe and North America. There were over 30,000 water wheels in the UK alone by the start of the twentieth century. By then, the most popular design was a horizontal axis *overshot wheel* whose form had incorporated the blades into the frame of the wheel, making them more effective at holding water and so more efficient for a given stream or watercourse. They are excellent where a feed stream is high enough so that it's above, or on a level with, the top of the wheel. If there is not enough height, then an undershot or *breastshot* would have to be used. Breastshot wheels basically use a lower water entry point, targeting the axle rather than the top or bottom of the wheel. They work out to be less efficient than an overshot wheel of the same size, but are appropriate where the terrain isn't particularly steep and where greater head is not readily available.

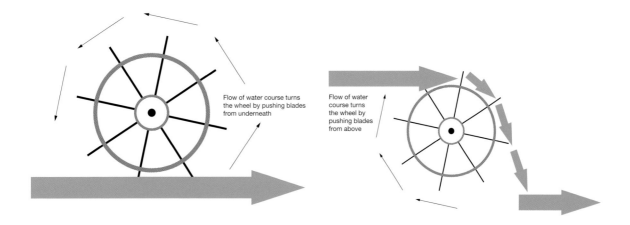

Figure 8.3 Undershot and overshot waterwheel designs

Source: Dilwyn Jenkins

The nineteenth century witnessed a massive take-up of mechanical hydropower, not just for use in milling grains but also to power bellows in iron furnaces and to lift water, minerals or other materials in mines and for industrial-scale processes. The first electric turbine appeared in France during the second quarter of the nineteenth century, leading to the initial peaking of hydropower uptake over the subsequent 100 years. Turbines were easier to find locations for than cumbersome large wheels, but more than this, in conjunction with better-made iron tubes and pipes in which water-flow was smoother, and under greater concentrations of pressure, turbines offered a vast improvement over the power efficiencies available from water wheels.

In the US at the start of the eighteenth century, mechanical hydropower was used extensively for milling and pumping. Two hundred years later, in the early 1900s, hydroelectric power accounted for more than 40 per cent of the United States' supply of electricity. During the 1940s hydropower provided around 75 per cent of all the electricity consumed in the West and Pacific Northwest, and about one-third of total US electrical demand. New innovations in other forms of electric power generation – from coal, oil, gas and nuclear – have meant that hydropower's percentage has slowly declined. Niagara Falls was the first US hydroelectric site developed for major generation and still supplies electricity today.

By the 1940s, before the national grid reached all of rural Britain, many of the towns and villages – particularly in the hills of Wales, Scotland and Yorkshire – were dependent on hydro electricity generators for power. The national grid, centralised coal power generation and, in the 1950s, the first generation of nuclear power plants led to the closure of most small and micro hydro plants in the UK. The remains of some can still be seen, many of them being considered industrial archaeological sites today. Principally the larger hydro schemes have been maintained in the UK and mostly large-scale or mega hydro projects have been built around the world since the 1950s. However, it should be noted that the high potential for adverse environmental and social impacts has meant that some mega hydro schemes are not considered as truly 'sustainable' by many in the industry.

Renewed interest in small and micro hydro has been evident since 1973, when oil prices first started to show unpredictability and an upward trend. Places like the Centre for Alternative Technology in Wales have pioneered its use and demonstration ever since the mid-1970s. With increasing GHG emissions targets at national and even organisational levels, plus fiscal incentives for clean electricity generation, over the first decade of the twenty-first century there has been renewed interest from households, farms, businesses, local authorities and governments in exploiting their hydrological resource for green power. This revived interest plus advances in technology – electronic controls, lightweight plastic tubing and better-made components – have created a modern hydro marketplace with cheaper, more reliable and more accessible equipment. With the arrival of Asian-made small and micro hydro turbines in the last ten years, the consequent drop in essential capital investment costs has enabled a whole range of new hydropower users to enter the market, many living in remote mountainous areas of the world.

Basic Physics of Micro-hydro Power

Water Flows, Catchment Areas and Available Head

The water that feeds a hydro scheme inevitably comes from somewhere above the intake. At the intake, the hydro system starts to take control of a portion of the water resource. Below, at the end of a pipeline (known as the penstock), lies the powerhouse, where the turbine is installed. As such, the typical location for a hydro installation is near the bottom of a hill or half way up or down a mountain. Apart from access to sufficient water and enough vertical height difference (head) between the intake and powerhouse, there are often other constraints such as access to the national grid or to buildings or settlements that could use the electricity generated. The cost of transmission cables for relatively short distances is generally not a major issue, but when transmission needs to be over miles or kilometres, cable length can make many potential schemes uneconomic.

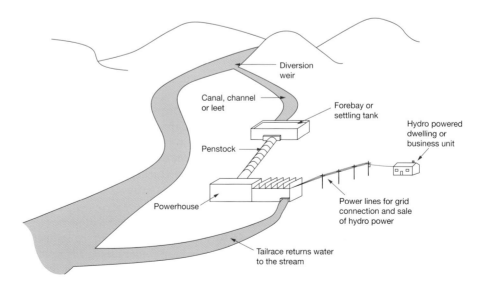

Figure 8.4 Common hydro system components

Source: Teilo Jenkins

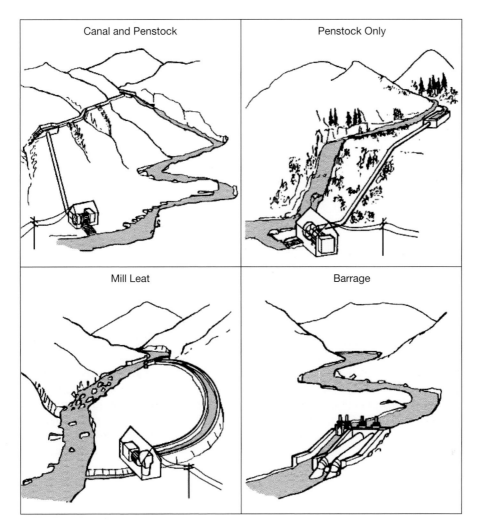

Figure 8.5 Common hydro system configurations

Source: British Hydropower Association, www.british-hydro.org

It's important to look at the whole catchment area for a particular water resource in order to understand how it might be maximised. The catchment area essentially comprises all the land above a hydro scheme's intake that feeds water into the watercourse. If it flows into another stream or valley, never reaching the watercourse identified for a hydro scheme, it belongs to a different watershed and catchment area. The catchment includes all the springs, bogs, lakes and streams that flow into the main watercourse above the level of the intake. Catchment areas form tree-like patterns on detailed contour maps, making it easy, once you know the site of the hydro scheme's intake, to mark off the relevant water catchment area.

Calculating the available head means choosing the location for both intake and powerhouse along a section of the identified watercourse. This inevitably involves a number of important considerations. Firstly, the higher up a catchment area the scheme is located, the less the total flow of water. So, even if there are amazingly steep cliffs offering wonderful head at the upper limits of the

Figure 8.6 Sketch showing water catchment area on contoured map in relation to intake, settling tank, penstock and powerhouse

Source: Dilwyn Jenkins

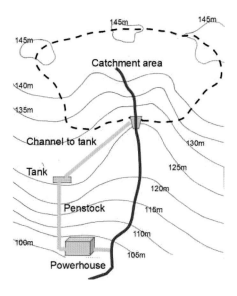

catchment, they may be of little use since their flow levels will be minimal. Secondly, the locations are influenced by terrain and topography both above and below the intake. The best site to place the intake and powerhouse along a useful section of the watercourse will usually be that which offers the steepest slopes and therefore usually the greatest head. Thirdly, there are financial considerations, largely in terms of the length and diameter of the penstock. Longer runs require bigger diameters, which in turn quickly increases the cost per metre.

Site selection is frequently complicated by a fourth set of factors: legalities of land-ownership rights and access to the watercourse as well as the best locations for intake and powerhouse. Once land rights and watercourse access issues are resolved or accounted for, the intake and powerhouse locations can then be selected according to the terrain, always looking for the biggest workable vertical height difference placed between them. This will be the available head. In ideal circumstances, the average gradient for a penstock on a micro hydro system should be as close as possible to 1:10 or steeper. For more detailed information on 'static' (or 'gross') and 'net' head, see the Desktop Survey section below.

Flows are also influenced by the permeability of the soil in the catchment area. Some soil types facilitate absorption while others tend to release precipitation into the stream much faster. In the latter case, it is certainly worth considering the advantages of some kind of reservoir to help retain part of the flow. This will allow the system to operate more smoothly between rainy periods. Making design decisions like this is where a professional hydro engineer or surveyor can be very useful.

In terms of making power calculations based on flows, it's useful to know the minimum flow figure and its average annual duration for the available watercourse (as opposed to simply the average or maximum) flows, since this will offer a good indication of how much power could be supplied by the scheme on a year-round basis. If too large a turbine is installed it may give full output at the rainiest time of year, but come the dry season it could well be too big to operate at all if only a fraction of the maximum flow is available at that time. Most micro hydro schemes aim to run at full power for around 50 per cent of the year, a figure that tries to optimise the kWh produced annually against the investment costs and likely available flows.

Calculating Hydro Electricity Outputs

The flow of a stream is measured in either litres per second (l/s) – and cubic metres per second (m²/s) – or feet per second (f/s). The electricity or power output of a turbine is usually measured in watts (W = joules per second) or kW (1,000 watts). Using these measurements, the standard hydropower calculation formula can be described as:

$P = Q \times H \times Y$ with the result in Watts

Where:

Q = the flow in l/s
H = the head in metres (m)
Y = the specific weight of water (9.81 kilonewtons/m²)
P = power output (kW)

Given this formula, a hydro scheme with a 0.03 m²/s flow and a 10 m head could in theory offer an output of almost 3 kW (or 2,940W): $0.03 \times 10 \times 9.8 = 2.94$ kW.

The purpose of a hydro turbine-system is to harness as much of this potential electricity as possible. However, there are several layers of efficiency losses: in the pipes (penstocks) and channels, as well as in the turbine itself and again within the generator. As a consequence, most schemes claim between 40 and 70 per cent overall efficiency (a good standard being around 50 per cent) in converting the potential hydro output into usable electrical power. So being more realistic, the example above would look like this: $0.03 \times 10 \times 9.8 \times 0.5 = 1.47$, where 0.5 is the 50 per cent reduction in efficiency.

However, since the energy produced is a function of the length of time the generator is running, it is also possible to use a formula for the number of watt-hours (W/h) a system will generate over a given period:

$E = Q \times H \times Y \times 0.5 \times h$ W/h

Where:

E = energy produced
h = number of hours generating

Most turbine controls incorporate a volt meter and an amp meter as basic components. Figures 8.7 and 8.8 show these meters for a 2.4 kW hydro system, one located at the powerhouse and the other inside the household that consumes the hydro electricity (the latter includes a total kW generated meter). If you

Figure 8.7 3 kW Pelton turbine at the Centre for Alternative Technology, Wales, UK

Source: Dilwyn Jenkins

Figure 8.8 Voltage, amps and total generation log at house which consumes the hydro electricity

Source: Dilwyn Jenkins

multiply the volts (300) by the amps (8) you can arrive at the watt output (2400 W or 2.4 kW).

Common Turbine Types

As with all hydro-electric systems, the idea is to drop the available water resource through a turbine that can transform its potential energy into first mechanical and then electrical power. There are two basic turbine types: impulse and reaction. Both types use a runner, or wheel, to convert the water-power into the power of a spinning shaft. Impulse turbines work best with higher head systems that feed the water down through pipes, while the latter is more suited to valley floors and rivers with little head but greater flows. Unlike impulse turbines, the reaction variety don't generally utilise jets for the pressurised and focused release of water, they just use the continuous flow of the watercourse itself against turbine runners. Impulse turbines are by far the most common type in small and micro hydro applications. As a rule they are also cheaper and smaller than reaction turbines. Small machines work for high head and low flow, while where the head is low a large flow is need and a larger machine to facilitate this flow. For the purposes of this book, we focus on impulse turbines.

There are a number of impulse-runner types designed differently to take as much energy as possible from the jet of water, the main varieties include:

- *Pelton wheels*: generally the most appropriate for micro hydro. A 5 kW turbine costs around £7,000 or US$11,000 upwards.

- *Turgo turbines*: very reliable but more expensive and best at heads of over 10 m.
- *Crossflow turbines*: lower maximum efficiencies but are cheaper since they're relatively easy to manufacture. A 5 kW installation costs about £9,000 or US$14,000 upwards.

The most popular reaction turbine used today is the Francis turbine. They require casting and are generally only used in hydro installations of 30 kW or larger.

How a Pelton Wheel Turns Water into Electricity

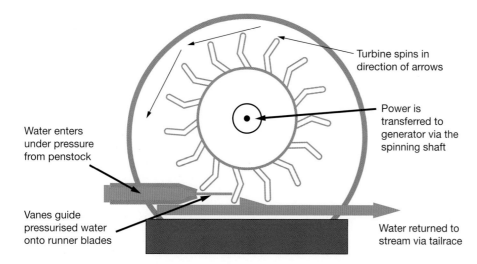

Figure 8.9 How a Pelton turbine works

Source: Dilwyn Jenkins

Figure 8.10 illustrates how each turbine has a slightly different way of delivering the water to the runners. The spinning of the runner turns the turbine shaft, which is fixed to the centre of the wheel. This is connected to the generator shaft. Gearings are usually incorporated to synchronise the relative optimum rotational speeds of the two. Greater efficiencies can be achieved with direct coupled drives, but this only works if the wheel and alternator are perfectly matched. Instead, belt drives are commonly utilised. For micro hydro systems, the electricity generated is usually single phase. Unlike wind and solar generating modules, micro hydro systems are designed to generate electricity AC (110/230 V), although small devices for charging battery banks also exist. There are two principal types of generators:

- Synchronous (alternator generation); and
- Asynchronous (induction generation).

Both alternators and induction generators offer different benefits and drawbacks. These should be discussed with the supplier.

Electronic load controllers (ELCs) have been a great boon to the hydro industry since they were first utilised just over 30 years ago. They have automated the

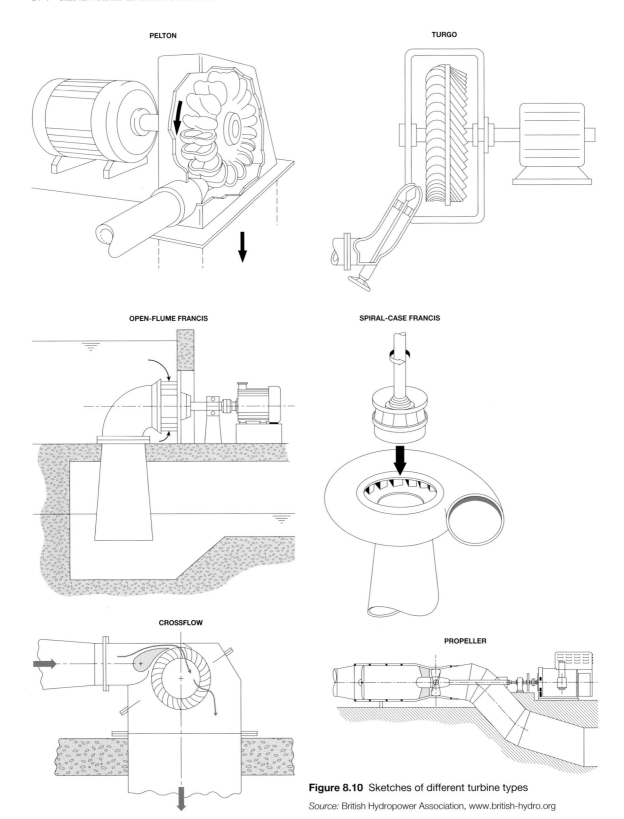

Figure 8.10 Sketches of different turbine types
Source: British Hydropower Association, www.british-hydro.org

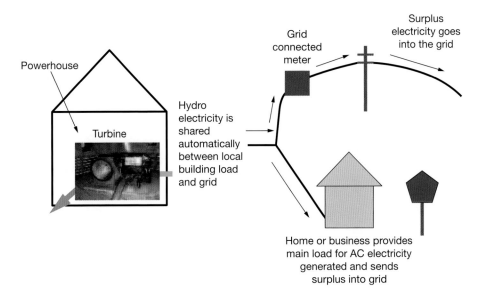

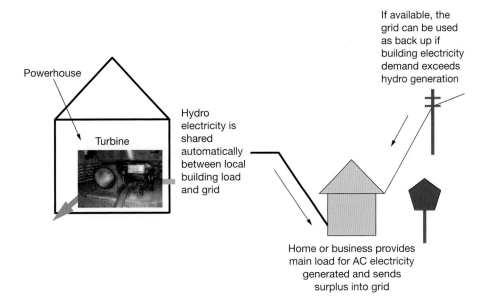

Figure 8.11
Common electrical configurations – generating for own use and/or for sale into the grid

Source: Dilwyn Jenkins

previous mechanical flow-control methods that used to govern the spinning speed of the alternator. ELCs can also be used to automate the switch-over between hydropower and mains power in a grid-connect situation. Once generated, the power has to be transmitted to the load or building via cables that are as short as possible to minimise cable efficiency losses. Armoured cable is the most robust and safest for external use.

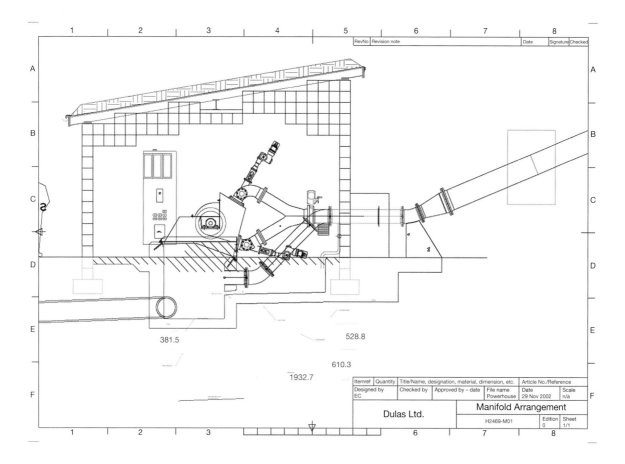

Figure 8.12 Design specifications in cross-section of a twin-Pelton hydro installation

Source: Dulas Ltd, www.dulas.org.uk

Desktop Survey and Initial Site Assessment

The two traditional phases of evaluating a potential hydro site are a desktop survey followed by an initial site-assessment visit. Detailed local maps (at least 1:25,000 with contours at least every 5 m) are required for the desktop study. With the right map and expertise it's possible to identify the catchment area and make some basic predictions about flow patterns, as well as mark one or two potential penstock runs where the steepest terrain allows.

Identifying Intake and Powerhouse Locations: Calculating the Available Head

Waterfalls can be very good for providing 'head', with the intake at the top and the powerhouse close to, but above, the plunge pool. However, perfectly positioned and accessible waterfalls are rarely available. Issues such as flooding, particularly for the powerhouse site, need to be considered. Look carefully at the surrounding landscape for evidence of flood plains, both regular low-lying floodplains and also (especially for locating the powerhouse) higher-level freak weather floodplains. Hydro systems are long-term investments, so it's an important precaution to protect them even against 50- or 100-year flood events.

Once the best potential intake and powerhouse locations are mapped, it should be possible to calculate the head. The accuracy levels of a map with only 5 m contours is likely to be very low for a micro hydro-sized scheme, so any map-based assumptions will need to be checked carefully at the field-visit stage. Basically, though, the gross or 'static' head is calculated from the vertical height difference between the intake and turbine location (i.e. the powerhouse). However, it is important to use a 'net' head figure for making power output calculations. Net head is gross head minus the pressure or head losses due to friction and turbulence in the penstock (which depend on type, diameter and length of the piping plus the number of bends or elbows). In this book we work with gross head for calculations, which is useful for roughly approximating power availability at a site. Net head, however, is a more accurate figure to use for calculating the actual power available. Calculating net head is beyond the scope of this book, but should be discussed with a hydro engineer before starting any installation.

Hydro sites can be broadly categorised as low- or high-head sites. Low head usually means a change in elevation of less than 3 m (10 ft), with high head being anything greater than this. Drops of less than 0.5 m (under 2 ft) are not really feasible for micro hydro schemes. A high flow-rate can compensate for low head, but the turbine will probably be significantly larger, less efficient and more costly than for a higher-head application.

Calculating the Flow

Making flow calculations is arguably more complex. The equation for available (annual daily average) flow, for instance, is:

Catchment area (m^2) × Annual run-off depth (m) / Number of seconds in year(s)

The catchment area can be calculated from the map work, but for reasonably accurate annual run-off depth figures – more or less equivalent to the average rainfall over the catchment area – the usual source are rainfall maps (also known as isohyetal maps). These are generally available from meteorological offices or, these days, online (e.g. in the US via http://wdr.water.usgs.gov/nwisgmap/). The number of seconds in a year is arrived at by multiplying the seconds in a minute, minutes in an hour, hours in a day and days in a year together ($60 \times 60 \times 24 \times 365 = 31{,}536{,}000$), but it is more frequently represented as 32×10^6. The idea of doing an initial full desk-study is to check the pre-feasibility of a project with a view to justifying the use of a specialist hydro consultant or land surveyor for a professional study and, if positive, a site visit. With some demonstration of feasibility, plus a rough idea of site-layout scenarios, the necessary tools are available to make sense of a visit to the site.

The key objectives of the site survey are to confirm, as far as possible, the head and flow calculations made in the desktop survey. These days, an accurate GPS altimeter reading is the easiest method for measuring the head from actual intake and powerhouse locations greater than 10 m head. Just ten years ago, more traditional methods such as theodolites, spirit levels, dumpy levels and staffs, water filled pipes and staffs or pressure gauges would have been more common.

Figure 8.13 Small hydro intake with Coanda screen (across centre of weir) and fish ladder (to right) on mountain stream

Source: Dulas Ltd, www.dulas.org.uk

If a nearby stream or valley has had a hydrological survey it may be possible to extrapolate useful data, correlating that stream's features with those of the site in question and making assumptions about them having similar run-offs, or adjustments if this clearly isn't the case. Alternatively, there is online software. For the UK, one of the best is Low Flow 2000 (www.hydrosolutions.co.uk), whose software suite was created to estimate river flows for any river reach within the UK, even where measured flow-data is not available. The website strongly recommends that LowFlows 2, the latest version, is used by competent hydrologists who have received appropriate training.

Since the available flow is hard to calculate with any real precision, only larger and more costly micro hydro projects are likely to justify a full hydrological study. These will gather all available data on rainfall from generalised maps and local weather stations over many years to the level of monthly information for the catchment area, then typically represent these as a hydrograph.

A standard rule of thumb for interpreting a hydrograph is to establish the maximum flow available for at least 50 per cent of the year. In the case of the example above this would be 250 l/s (equivalent to 530 ft^3/s), since this is the largest flow available for more than seven months of the year. If you wanted to keep capital costs down, you might go for the 200 l/s (equivalent to 423 ft^3/s) figure which is available for ten months (83 per cent) of the year. This information can be used to calculate the likely annual electricity output of a proposed scheme with some accuracy. For greater accuracy still, an onsite flow monitoring exercise can be invaluable.

Commercial installations might fix an automated stream-gauge and data-logger at the site to take readings at 15 or 30 minute intervals. Less accurate but still scientifically valid, the velocity-area method is another option. Since flow at the point of intake is essentially a relationship between the volume and the speed of the water flowing past, if we can work out the speed (velocity) and also the cross-sectional area (a snap-shot of the volume), then simply multiplying these together will give us a flow rate. Calculating the cross-sectional area can be done quite simply, as illustrated in Figure 8.15, by taking several depth measurements

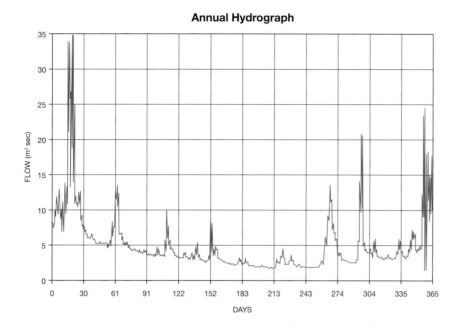

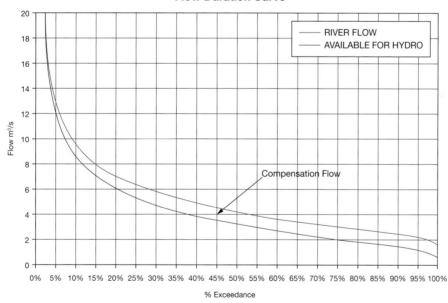

Figure 8.14 A typical hydrograph for a micro hydro site shows available flow in cubic metres per second (sometimes as litres per second for smaller systems)

Source: British Hydropower Association, www.british-hydro.org

(A to J, in this example) spread equally apart across the stream width and then taking the mean average of these measurements (e.g. add up all the depth measurements then divide this total by the number of depth measurements). This average should be multiplied against the width of the stream where the depth measurements are taken. The resulting figure is the cross-sectional area, or a good estimate of it.

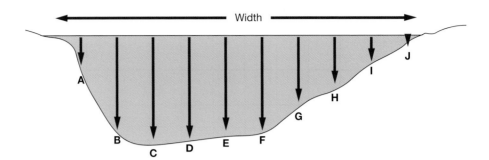

Figure 8.15
Calculating the cross-section of a stream

Source: Dilwyn Jenkins

This relies on making calculations for a stream's cross-section at a point where it is possible – using a current meter – to measure the speed of the current. By multiplying the current against the cross-sectional area (m^2) by the velocity of the current (m/s) you can arrive at the flow rate. If an electronic current meter isn't available it's possible to estimate a stream's current velocity by timing the movement of a float between two measured points along the course.

There are a variety of low-tech ways to measure flow, of which the bucket method is arguably the simplest and most workable for smaller streams. This involves filling a container of known volume with the entire stream's flow, often requiring the building of a small dam and removing the flow from the dam in a pipe which can be directed into the bucket or measured recipient. This is usually done at least three times and then the 'bucket' volume is divided by the average time to give the flow: Flow (l/s) = volume of water (l) / time to fill (s). By repeating this exercise weekly, over one or more years, it is possible to get quite accurate annual available-flow assessments for small streams.

Micro hydro systems between 500 W and 5 kW

At the lower end of the micro hydro scale (500 W to 3 kW), most systems are used for battery charging, where the energy will be stored as DC. The battery bank is generally linked to an inverter, which can transform the electricity back to the correct AC voltage for compatibility with standard household or business appliances. In most circumstances, hydro battery-charging systems will be associated with off-grid buildings. They could also be used in grid-connected situations, although where the grid is available it is simpler and more efficient to use the AC generated electricity directly from the turbine. Any surplus could be exported to the grid or it would have to be dumped (usually into hot water tanks).

Figure 8.16 shows two examples of micro hydro turbines from opposite ends of the scale. The US-made Harris micro hydro generator is a small scale, compact and efficient battery-charging Pelton turbine which utilises a cast bronze Pelton wheel in a cast aluminium housing. It is designed for high head (20 to 600 feet), low flow (4 to 250 gpm) applications and uses Ford automotive alternators, keeping the price reasonable and making spare parts available from your local auto parts store. The DC power generated by the micro hydro turbine steadily charges your battery bank (12, 24 or 48 V) for future use. These small systems

Figure 8.16a The Harris Compact Micro-hydro
Source: Harris Hydro

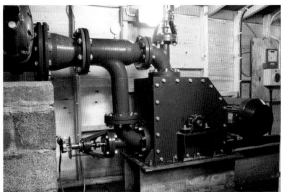
Figure 8.16b High head Pelton turbine (5.5 kW)
Source: Turbine Services, www.turbineservices.co.uk

operate 24 hours a day collecting energy, a little at a time, to be delivered 'on demand' from the batteries as the need dictates. An inverter can be used to convert the DC electricity to household AC power, and a diversion load is used to burn off excess power when the battery bank is full. The 12 V model can produce up to 700 W, the 24 V models can produce up to 1400 W, and the 48 V models can produce up to 2,500 W of power, depending on the site's head and flow rates. Sites with higher head are most desirable because they need less water, smaller piping and fewer nozzles, cost less to install and fare better in low-water years. If output is more than 500 W the PM brushless alternator should have the optional Fan Kit.

A much large system, the Pelton Turbine is working off 100 feet of head and has two valves/ nozzles. Both are controlled using motorised spear valves. As the river dries up in the summer months the valves wind in to maintain power-production in varying water conditions. The turbine produces 5.5 kW and connects directly to the mains electricity supply.

Micro Hydro System Design

Relatively high installation costs, the large amount of civil works sometimes required and the safety issues surrounding high-pressure water in pipes all combine to emphasise how important it is to design a micro hydro project in a detailed way so that it is as effective as possible. Apart from the powerhouse the main tasks include civil works, including the intake and any channel to the settling tank and the all important penstock, which acts much like the barrel of a gun, concentrating the force in one direction. In the case of a standard Pelton wheel hydro the direction is toward the vanes, which spray the pressurised water onto the spoons or blades of the wheel gaining force from the available head, right at the end of the penstock.

Intakes and Diversion Weirs

A diversion weir is fairly standard for an intake, stabilising the water and diverting through a stop log or sluice gate a part of the flow into a usually open channel, before entering the settling tank. Weirs vary in size, height and complexity according to the scale of the micro hydro system as well as the size and run-off conditions of the stream. Water should be able to flow naturally into the channel via the sluice gate or a compensatory-flow notch, which should be located carefully above the stream-bed level to ensure that maintenance of the design flow for the stream can occur all year round. This guarantees that the hydro never steals too much water from the stream, something which could cause permanent ecosystem damage in sensitive or special environments.

The series of photos below (Figure 8.17, 8.19a, 8.19b, 8.20a, 8.20b, 8.23a, 8.23b, 8.24) were taken at the site of a 2.4 kW micro hydro turbine which has been operational on an average sized stream for over 30 years. The weir is made of shuttered concrete poured, in two halves, over boulder ballast. In this example the weir was renovated and redesigned about 18 years ago when a new, more innovative intake screen and a new penstock were incorporated. The new screen-based intake is on the far side of the weir. The U shaped groove built into the weir can be opened (by lifting the simple wooden sluice gate) in flood conditions to relieve pressure on the weir itself.

There are various techniques for diverting some of the watercourse into the penstock. Many involve the use of a screen to capture any silt, sand and minute gravel, which would damage the turbine if it travelled down the penstock. As far as possible, it's good to work with the natural features and flow of the watercourse: not only does this keep civil works to a minimum, but it should also make

Figure 8.17
Watercourse and diversion weir from above intake

Source: Dilwyn Jenkins

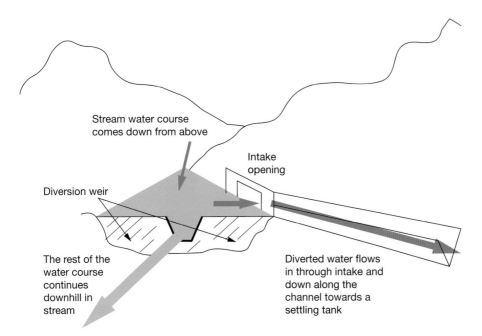

Figure 8.18 Intake and diversion weir diagram

Source: Dilwyn Jenkins

for a stronger, long-lasting intake. Straight sections of the watercourse are best for intake location. Alternatively, the outside of a bend is recommended.

Some intakes make use of a side channel, usually with its own sluice, that carries the water into a settling tank where particles will tend to sink or, again, be trapped by some kind of meshed screen before entering the penstock. Settling tanks should have sufficient depth and space below the main flow of the water so that any particles can sink out of the full force of the water as it heads towards the penstock. Settling tanks will need regular cleaning out: the frequency will depend on the season, the turbidity of the water (the degree to which it contains suspended solids) and the tank's depth relative to the main watercourse flow across it. Channels are sometimes simpler and cheaper than the penstock, so their use in moving water to the top end of the penstock can help cut overall installation expense.

Among the best screens available on the market are those that utilise a fine horizontal bar design in order to take advantage of the coanda effect (the tendency of a fluid jet to be attracted to a nearby surface), stopping particles entering and self-cleaning leaf debris while at the same time allowing enough water to pass through into the intake and penstock run. Non-metal screens are advisable for regions with severe winter freezing conditions.

The old penstock for this hydro site (Figure 8.19b) can be seen on the left-hand side of the photograph, just below the flood-relief notch. The new intake utilises a coanda effect screen and the new penstock can be seen heading downstream just below the base of the screen. The penstock then heads downhill and (in this case) for the first part of its course, along the stream bed itself.

Figure 8.19a Intake and diversion weir

Source: Dilwyn Jenkins

Figure 8.19b Old and new intakes at diversion weir

Source: Dilwyn Jenkins

Source: Matt Palmer, Dulas, www.dulas.org.uk

The Penstock

The scheme in these photographs doesn't need a settling tank because a very effective screen has been installed at the intake. Ideally, penstocks run downhill as steeply and directly as the terrain and other factors permit, straight to the powerhouse.

Penstock design is important, not just because it represents a large proportion of the scheme's capital costs, but also because of its relevance to maximising the available head, which is effectively the same as maximising the installation's power output. An inadequate penstock might not last long due to pipe deterioration or breaking under the high pressures demanded of it. The 'static' pressure

experienced by the penstock increases in relation to the head by one bar (100 kPa) for every 10 m, but there's also a 'surge' pressure to consider.

Surges in pressure, often caused by temporary pipe blockages, can increase penstock pressure by more than another 50 per cent, so this needs to be added to the static pressure calculation. Due to the pressures they work under, penstock pipes will tend to move over time, particularly towards the outside of any bends. Given this, penstocks are usually anchored at appropriate points, with particular reinforcement at the outside of bends. The anchors can be metal or wooden stakes, although they are frequently made of concrete embedded into the ground. At the bottom of the penstock, usually inside the powerhouse or just outside, most hydro schemes incorporate a gate valve, which permits turning the water off from the pelton – for any wheel, belt drive or turbine maintenance – without emptying the penstock itself.

Figure 8.20a First section of penstock

Source: Dilwyn Jenkins

Figure 8.20b Simple hole valve in penstock

Source: Dilwyn Jenkins

Figure 8.21 The turbine house at the community hydro project in Talybont-on-Usk, UK. The intake is at the reservoir further up-river. The owners of the reservoir, Welsh Water, are obliged to maintain the flow in the river and some of this 'compensation' flow is channelled through the turbine.

Source: Talybont-on-Usk Community Hydro, www.talybontenergy.co.uk

Most micro hydro systems will be able to use the lighter-weight PVC-U plastic pipes for the penstock, which tend to be smoother internally than steel or iron pipes, making for less potential head loss from friction as the water flows down. This must be buried or covered where possible to protect against degradation from sunlight. While certainly cheaper, PVC-U is not very environmentally friendly. Marginally more expensive, yet more resistant to sunlight, polyetherine is a good alternative piping material. Steel and cast iron or ductile iron options are harder to get to site and considerably more expensive materials that last longer than plastics in an over-ground penstock scenario. The diameter of the pipe is usually calculated in relation to the ideal design flow-rate, given the available flow and selected turbine design and size. Any experienced hydro engineer will be able to advise on the best pipes available in your region for different pressures of penstock.

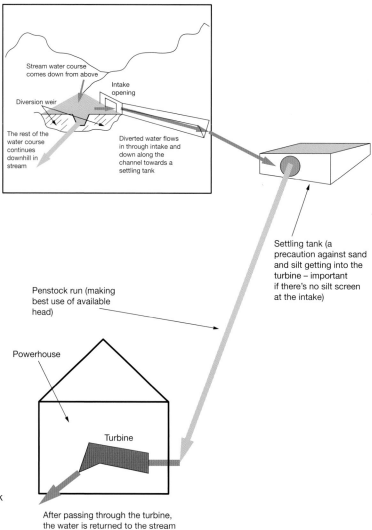

Figure 8.22 Settling tank, penstock and powerhouse diagram

Source: Dilwyn Jenkins

HYDROPOWER 187

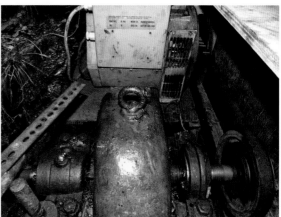

The Powerhouse

Guided by hard metal jets onto the Pelton wheel's spoons, the pressurised water from the bottom of the penstock spins the wheel, often at well over 1,000 revolutions per minute. The Pelton wheel's axis is linked to a drive belt that, in turn, spins the turbine which generates electricity. Meanwhile, the water that passes through the intake, penstock and Pelton wheel is returned to the stream by what is usually termed a 'tail-race' that is connected to the powerhouse. Sometimes it is piped or channelled out of the powerhouse back to the stream.

In this example the tail-race simply spills back into the stream (at the centre of Figure 8.24) from a simple but adequate powerhouse. The example in the

Figure 8.23a
Penstock arrives at powerhouse
Source: Dilwyn Jenkins

Figure 8.23b Pelton wheel, drive-belt and turbine
Source: Dilwyn Jenkins

Figure 8.24
Powerhouse with water (centre of picture) returned to stream
Source: Dilwyn Jenkins

photograph is open-sided but well hidden from public view and difficult to access. Many powerhouses will need to be secure against theft or vandalism. Enclosed powerhouses will need to consider a balance between natural light for daytime visits and an electric bulb for unforeseen night-time emergencies. The vagaries of local flooding and the annual weather regime will need to be considered. In many scenic areas, there will be local-government building regulations impacting on the architectural style of the building. If there are neighbours close by and noise is an issue, the powerhouse should be enclosed and well insulated.

Typical Applications for Homes and Businesses

Why would a household or business that is already linked to the grid take the trouble to invest in a micro hydro scheme? The answer, these days, is simple economics. With rising energy and electricity prices, and increasingly available preferential tariffs associated with the generation of clean and green power, a reasonably good micro hydro site will pay for itself in less than ten years. After that, the scheme will provide free (after maintenance and minor repair costs) electricity for at least another 40 years, and green preferential tariffs for as long as this is one of the ways in which governments try to minimise climate change. If a building that needs powering actually has the necessary water resource, along with access to capital and preferential tariffs, whether it's a home office or workshop, it makes business sense to invest in a micro hydro scheme.

The Grid-connected Hydropowered Home

Even in a situation where a house is already connected to the grid, there are a number of advantages to being connected:

- offsetting electricity (very useful given the high prices today);
- selling clean electricity (e.g. presently in the UK, hydro electricity can be sold to the grid at over £0.20p per kWh on a 20 year contract); and
- matching demand loads of home.

The Grid-connected Hydropowered Business

While the advantages of having hydropower for a grid-connected business are similar to those for a home, many businesses may be able to gain more, financially and environmentally, partly through sheer scale of use but also through the design of its work schedule to optimise demand for energy at times when most hydropower is available:

- offsetting electricity (useful even for businesses);
- selling clean electricity; and
- matching demand loads of business (e.g. mechanical, refrigeration or heat processes).

Incentives for Hydro Business and Home Investors in the US

In the US most states have some incentives that can be applied to micro hydro systems, although it's fair to say that they are more plentiful for small wind and solar systems. The best place to start is the DSIRE database (www.dsireusa.org). Some 14 US states have established clean energy funds to promote renewable energy and clean energy technologies. Direct Production Incentives, similar to a production tax credit, can provide monetary income. At the Federal level, Section 1212 of Energy Policy Act of 1992 (EPACT) provides a Renewable Energy Production Incentive (REPI) of 1.5 cents per kWh to non-profit organisations that own micro hydro energy facilities. On top of this, Property Tax Reductions promote renewable energy development by decreasing the tax burden associated with owning a micro hydro facility, something that is relatively high compared to fossil energy because of the greater land requirements for hydro systems per unit of output. Accelerated depreciation is usually permitted for hydro capital investment.

Business-owned or community hydro schemes might consider green marketing of the clean electricity. A business on an industrial park could, for instance, sell the electricity to its other commercial neighbours. The price could be competitive and it would bring green prestige to the estate. These are usually voluntary programs in which customers agree to pay a premium to purchase 'environmentally friendly' or 'green' electricity. This encourages the development of a market for renewable power. Net metering is new in the US, but it has started and this system gives utility customers a guaranteed market for their hydropower by operating a 'reversible meter'. When customers use more electricity than they generate, they pay for the additional electricity at retail prices as usual. When they generate more electricity than they use, the electric utility is obliged to purchase the additional electricity (although the prices received varies widely by state and region).

Community Hydro

Community hydro projects are likely to be in the small hydro range (100 kW to 5 MW) rather than micro hydro (up to 5 kW). The logic of planning a community hydro scheme is similar to domestic installations, and both domestic and commercial hydro involve similar risks that demand the involvement of a professional or experienced hydro engineer. As with many other types of community initiatives, there are a number of additional issues that need to be addressed:

- the model for developing the project (e.g. a developer-led initiative with community investment and benefit, or perhaps a community-led initiative with bought-in expertise);
- typical barriers to community projects, such as the need for a very determined core development group, access to finance and access to specialist knowledge; and
- legal structures: some body or organisation has to own the plant and be ultimately responsible for it. There are various models appropriate for community ownership, including a Company Limited by Shares, a Company

Limited by Guarantee, and an Industrial and Provident Society. Or a Registered Charity could be set up.

Regulations and Incentives

The regulation of hydro schemes in the US is covered at both federal and state levels, but the first point of contact for information on restrictions around water resources in any given locality should be the county engineer. There are two essential and one recommended federal agencies that should also be consulted in the early stages of planning a micro hydro scheme (contact details can be found in Annex 2):

- The Federal Energy Regulatory Commission (FERC): essential consultation;
- The US Army Corps of Engineers: essential consultation; and
- The US Fish and Wildlife Service: recommended consultation.

In the US the Public Utility Regulatory Policies Act (PURPA, 1978) requires electricity utilities to purchase power from independent power producers if certain basic conditions are met. The local utility and/or public utility commission can advise as to precisely what technical and operating requirements are relevant in your area. They should also be able to give a price that the utility will pay for the clean electricity generated. Licences may also be required from FERC, and most connection requirements insist on basic insurance cover that, in some circumstances, may cost more than is earned from selling the power. Tax credits are available in some US states to businesses that invest in hydroelectric systems: most businesses qualify for such credit (usually up to 35 per cent of the project cost).

Regulations tend to be tighter in the UK, where every hydro scheme has to apply to the Environment Agency for a water abstraction licence. This often involves the need for an environmental impact assessment (EIA) and will determine the maximum flow of water the scheme can remove (even if temporarily) from the watercourse. In the UK this is often as little as 25 per cent or less of the minimum flow to help maintain healthy aquatic biodiversity. Screens to protect the fish from entering an intake and sometimes even fish-ladders to avoid inhibiting fish breeding may be requested as a conditions of a licence. Well-designed micro hydro projects should not have a negative impact on fish life in the stream or river. Works-in-river consent may also be needed from the Environment Agency. For new hydro schemes, planning permission is also required from the local authority for the whole scheme, from intake to powerhouse.

Deciding on Hydro

Most of the main issues are summarised in Table 8.1, divided into the physical aspects of the resource on the one hand and the financial scenario on the other. If you can tick 'Yes' to every criteria question then it is likely that a hydro turbine is a good choice for your needs. The most important criteria relate to the water resource, although finance and the potential for income generation or cost-offsetting are also very important factors.

Table 8.1 Critical criteria for deciding whether hydropower is suitable at a specific site

Critical criteria		Yes	No
The resource	Is there sufficient flow of water in the stream?	☐	☐
	Is the flow enough to keep generating electricity even during the driest time of year?	☐	☐
	Is there enough gross head to generate the required electricity?	☐	☐
	Is the water resource close enough to the building or grid connection?	☐	☐
	Are legal rights of access to land and water established or possible to establish cost-effectively?	☐	☐
Financial issues	Is a preferential tariff available for the sale of electricity from the site?		
	Are there any capital grants available to help cover the installation costs?	☐	☐
	Are there any tax incentives for construction or investment?	☐	☐
	Can the installation meet all the requirement conditions associated with tariffs, grants and/or tax incentives?		

Off-grid micro hydro

The lure of having access to essentially free electricity from a micro hydro scheme for 50 years or more is very clear in a situation where a house or enterprise is located off grid but requires power. With off-grid micro hydro the hydro electricity arrives via cable to a battery bank connected to an inverter that transforms the power into 110/230 V AC. The batteries are kept safely in an airy wooden structure that can be closed against unwanted interference and kept safe from accidental damage or spillage.

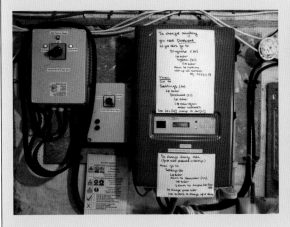

Figure 8.25a Hydro inverter controls located in premises being powered to enable remote monitoring

Source: Dilwyn Jenkins

Figure 8.25b Battery bank stored in airy wooden cupboard close to inverter controls

Source: Dilwyn Jenkins

9
Case Studies

Energy Efficiency and PV Power for Business Unit

Case Study 1: Low Energy Business Unit with
Grid-connected PV array, Dyfi Eco Park, Wales, UK

Unit 1 at the Dyfi Eco Park in mid-Wales, UK, has been rented by the fast growing renewable energy company Dulas Ltd. since it became available in 1997, and makes an interesting case study that combines elements of renewable power inputs with a general low-energy design.

It is built largely from local Welsh semi-durable softwoods such as Douglas fir and larch. As far as possible other materials were sought locally too, in order to keep the embodied energy of the building as low as possible. The elimination of individual steel portals and any associated massive concrete footings also saved significantly on the building's eventual embodied energy. Where possible, the wooden fabric of the building was used to provide structural integrity. Using a

Figure 9.1 A low-energy business unit with own PV array, built in 1996 (UK)

Source: Dulas Ltd

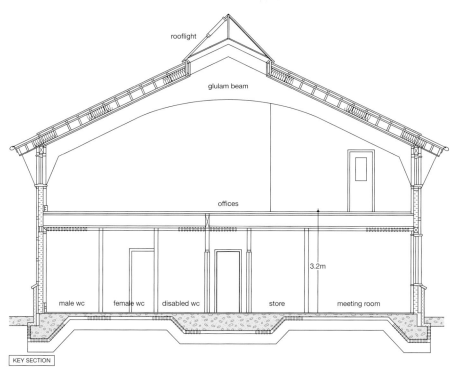

Figures 9.2a and 9.2b A photograph showing the use of timber in Unit 1, Dyfi Eco Park business unit, and a sketch which illustrates well the design for the apex of the building to allow for solar gain

Source: Dulas Ltd

timber-framed deep panel design meant that there was plenty of room in the stud compartments to fill with cellulose-fibre insulation.

Much of the building's charm as well as energy design rests in the glazing, which allows ample solar gain into the unit, particularly to the open-plan upstairs section. One of the few faults found by the building's users is that the upper storey has too much sunlight and, in the summer, too much space heat from solar gain. The glare of sunlight made it difficult for those upstairs to view their computer screens, and as a consequence the occupants compensated by placing a calico cotton cloth along the inside of the glazed apex, diffusing the bright glares from sunlight during the working day.

Figure 9.3a and 9.3b
Photographs of interior upstairs of Unit 1, showing the solar glare from skylight apex and also the cotton sheet used to resolve the problems caused by glare

Source: Dilwyn Jenkins

The building also has high levels of draught-proofing and natural ventilation. After 15 years, some of the double-glazed window units are suffering from internal condensation, but there have been no specific signs that the air-tightness of the building has worsened significantly over time. A glazed gable to the southeast performs an important solar gain function but, again, a curtain has been installed by the occupants to stop the stronger sunlight and additional heat on summer mornings. This gable is also screened (sun shaded) by three tiers of external timber slats and, in recent years, some deciduous trees.

The property and its water are heated by natural gas using a small domestic-sized condensing boiler. At the time of building a separate gas pipe was run under the nearby main road to facilitate potential future connections to a local methane producer. The unit had extremely good air change per hour test results (1.5 per hour). Many average buildings suffer from 8 to 12 air changes an hour with all controllable openings closed: the Canadian standard is 2.5 and a good British standard would be around 4.5.

Soon after taking over Unit 1, Dulas decided (as a company that installed PV) that they should have their own PV array. This was designed and built to provide around 10 per cent of the building's electricity use and has generated 7,820 kWh between 1998 and 2011, offsetting 3.3 metric tons of CO_2 to date. Electricity consumption has grown faster than predicted along with expansions in staff, so the 10 per cent figure has not been achieved, but there are plans to install more PV at the site.

Figure 9.4
Photograph of PV array built over a bicycle shed at the front south-facing corner of Unit 1

Source: Dilwyn Jenkins

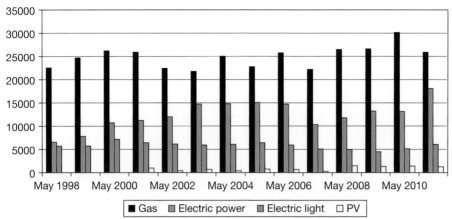

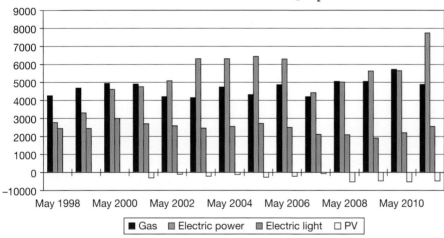

Figures 9.5a and 9.5b
(a) annual energy consumption and (b) annual CO_2 emissions for Dulas Unit 1 (1998 to 2010). The years with heaviest gas use had colder winters and the increasing trend for more electricity use is do to with greater staff numbers using more and more gizmos that need plugging in or charging (except for a dip in 2006 to 2008 when staff were moving out as the company acquired an additional business unit on the same park).

Source: Dulas Ltd, www.dulas.org.uk

Energy Efficiency and Wood Pellet Heating for New Build Home

Case Study 2: Pellet Central Heating Boiler with Solar-assisted Domestic Hot Water in Low Energy New-build House, Wales, UK

This example depicts a super-insulated new-build detached house in temperate climatic conditions located in rural Wales, UK. It is a wooden-framed building with three bedrooms and a larger open-plan ground floor with living spaces and kitchen. Domestic hot water is assisted by a small flat-plate solar collector positioned on a south-facing roof.

Located in the centre of a village, the house is not particularly exposed to local weather conditions. The windows and doors are triple glazed. The roof and floors

Figure 9.6
Photograph of low-energy house with pellet boiler and solar array

Source: Dilwyn Jenkins

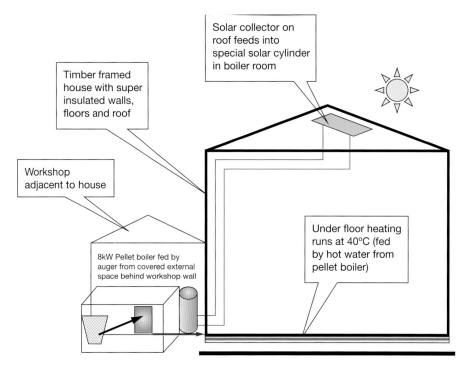

Figure 9.7 Sketch depicting low-energy house with pellet boiler and solar array

Source: Dilwyn Jenkins

are insulated significantly more than the recommended minimum (Part L of the UK Building Regulations). Exterior walls contain 145 mm of solid foam insulation and there are plans to install a heat recovery system which would make the property even more energy efficient.

The concept behind this boiler installation was to minimise pellet fuel-consumption by maximising both thermal and combustion efficiencies. An 8.2 kW boiler was installed in a garage-sized outbuilding adjacent to the house. The same outbuilding houses a solar hot-water cylinder and pump stations. A twin-walled stainless steel flue was installed in the boiler room. The heat distribution is based on a multi-zoned underfloor coiled system which operates at only 40°C for the ground floor, and upstairs there are steel wall-mounted radiators, each with independent thermostatic valves. There are 4 m^2 of flat-plate solar collectors, selected for compatibility with the pellet boiler and automatic control system.

The pellet store is a flexible tank located directly behind the boiler, often just the other side of the external wall in a separate covered space that can be filled from an open yard belonging to the adjacent property. Wood pellets are fed by screw auger through the external wall into the boiler's combustion area. To date the pellets have been delivered in bags and the flexible tank has been filled by hand, partly to keep a check on the quality of the wood pellets in the first year of operation. Bulk pumped tanker-deliveries are, however, possible.

According to the property owner, the 8.2 kW pellet boiler would be capable of heating a significantly larger space, given the same very high levels of insulation and similar climatic conditions and building exposure levels. The house is warm all year and the boiler only fires up a few times a day, ensuring low levels of pellet fuel consumption. After the first winter, some of the initial three tons of

Figure 9.8
Photograph of mixing system and pipes for underfloor heat distribution

Source: Dilwyn Jenkins

pellet fuel still remained unused in the store. It is quite possible that a smaller boiler could effectively heat this thermally efficient house.

Solar Water Heating

Case Study 3: Apartment Complex with Solar Hot Water, California, US

This case study demonstrates how California apartment building businesses can benefit substantially from solar water-heating. As in many US states, apartment building landlords in California tend to include hot water with every rental lease and utilise central water-heaters (usually gas fuelled) to provide hot water to residential units and on-site communal laundry facilities. This means that every bath, shower, dish washing or laundry load costs the landlord money.

In this example, the apartment or condo complex has around 120 residential units and 160 residents. For each unit the calculation is as follows: 20 gallons per person per-day for the first person, 15 gallons per-day for the second and 10 gallons per-day for each person thereafter. Also included in the calculation are the 12 front-loaded energy-saving washers to a total hot-water consumption of 3,000 gallons per day. The Californian company Solar Thermal Solutions provide this case study data on their website (www.freehotwater.com).

The minimum daily heat demand at 80 per cent efficiency is reported to be 1,875,150 BTU. The system for this apartment block has 3,000 gallons of water storage. Before installing 66 solar collectors and taking up 18.6 m^2 (3,500 ft^2) of roof space, the gas hot-water heating bills were calculated at being around $16,000 a year. The gross installation costs came to $180,000, but the project (executed in 2010) received an $86,000 state rebate on the solar installation and a $54,000 (30 per cent) US Federal tax credit, so the net system cost was effectively just $40,000. Based on the net costs, the system paid back the capital investment in three years. From the fourth year on it will continue to save the equivalent of around $16,000 a year.

Case Study 4: Pellet Central Heating Boiler with Solar Assisted Domestic Hot Water in Standard House, Wales, UK

This example depicts an average sized (three bedroom) detached house in temperate climatic conditions located in rural Wales, UK. Domestic hot water is assisted by a small solar thermal collector positioned on a south-facing roof. There are two adult occupants with occasional visits by relatives and friends.

Although an older property, insulation measures have been installed in the ceilings, roof and floors. The roof has 300 mm of mineral fibre insulation (reduced to 75 mm around the sloping ceilings). Underneath the solid floors there is a 30mm layer of foam insulation. The property is double-glazed (K-glass type). There is a large pre-existing chimney in the house that produces some air loss. The previous heating system was comprised of electric storage heaters augmented by a wood stove in the main living room. In the depths of winter the occupants were rarely as warm and comfortable as they wanted to be.

CASE STUDIES   201

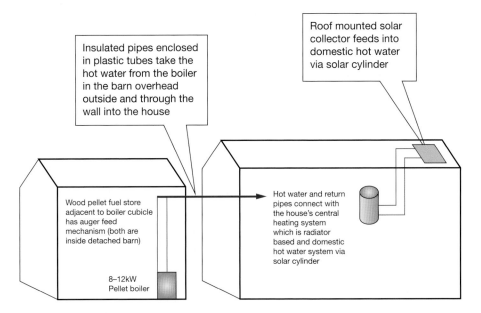

Figure 9.9 Sketch depicting wood pellet and solar water heating working in combination, Case Study 4

Source: Dilwyn Jenkins

Figure 9.10 Photograph of standard house with pellet boiler and solar array

Source: Dilwyn Jenkins

The boiler installed is rated at 15 kW and has been sited adjacent to a flexible tank pellet-store inside a barn just a few metres from the house. A screw auger feeds pellets from the store into the combustion area of the boiler and the hot water produced is delivered by well-insulated overhead copper pipes from the boiler into the house where it connects with a heat-distribution system which utilises standard wall-mounted steel radiators with independent thermostatic valves. No accumulator was installed. A twin-wall stainless steel flue was installed in the boiler room. The occupants are now warmer and more comfortable in the winter months and there is high customer satisfaction.

The boiler has a summer setting for domestic hot water which fires up twice a day to suit the occupants' lifestyles. If the weather is particularly sunny and domestic hot water needs are met by the evacuated tube solar thermal array, then the boiler will not fire. Domestic hot water is heated in a twin-coil solar cylinder.

Wood pellets were originally sourced locally but were not of the quality required. Better quality pellets are now bought from a new pellet manufacturer located about 80 miles away. Deliveries are in bulk by tanker, approximately every nine months. The property consumes between five and six tons of pellets annually.

Case Study 5: Domestic Solar Water Heating, Sussex, UK

A retired couple in Sussex, UK, recently downsized to a three-bedroom bungalow, which has been retrofitted with both a solar thermal and solar photovoltaic (PV) system. They currently benefit from feed-in tariff (FiT) payments derived from their PV installation. One of their main reasons for downsizing and installing renewable energy technology was to minimise their financial outgoings, and in order to maximise their savings they have also added additional loft and cavity wall insulation. The key factors driving them to invest in a solar thermal system (as well as PV) were a lack of return from traditional investments in banks and increasing energy prices. They were also looking towards retirement when their monthly outgoings would need to be reduced.

The cost of the solar thermal installation was approximately £4,000, of which they received grant assistance of £400 through the Low Carbon Building Programme. The rest of the money was raised through personal savings. The couple feel that the expectation of reducing their energy bills have largely been met, particularly in the summer months where they believe the solar thermal system can provide up to 100 per cent of their hot water needs. The installation and operation of both the solar thermal and PV system has been very pleasurable with immediate benefits in terms of financial savings. They say that they would definitely recommend both solar technologies to others and would also consider an additional renewable energy system such as a heat pump (information sources from EST, www.energysavingtrust.co.uk).

Heat Pumps

Case Study 6: Changing Heating System from LPG to Air Source Heat Pump, Gloucestershire, UK

Ecovision Systems (www.ecovisionsystems.co.uk) installed an air source heat pump and new wet heat-distribution radiator system to reduce liquefied petroleum gas (LPG) heating for the provision of hot water and space heating in a standard semi-detached house in Gloucestershire, UK. The new system runs at less than 33 per cent of the original fuel cost and the householders benefit from

Figure 9.11a, 9.11b and 9.11c Photographs of a 14 kW Ecodan air source heat pump (ASHP) installed in a Gloucestershire village, UK

Source: Ecovision Systems, www.ecovisionsystems.co.uk

reduced fuel bills, reduced carbon emissions and the knowledge that a renewable energy installation can enhance the value of the property.

The Gloucestershire village in question was not on mains gas, which made the economics more favourable for a heat pump installation. A 14 kW Ecodan air source heat pump was installed. Other householders in the village are also looking to replace their LPG or electric storage-heating systems for heat pump systems.

Case Study 7: Domestic Ground Source Heat Pump system, Wales, UK

A barn conversion in mid-Wales, located next door to the main house, turned a dirt floored, open barn to a dwelling in 2008. Since they were already implementing such significant works the owners decided to go for underfloor heating with a heat pump supplied by John Cantor Heat Pumps. They selected a Stiebel Eltron 1.5 kW heat pump with a horizontal ground source system. The heat pump

Figure 9.12a, 9.12b and 9.12c Various perspectives on a barn conversion with horizontal ground source heat pump system, showing (a) the house itself, (b) the external housing for heat pump and (c) the internally located manifolds, Wales, UK

Source: Emma and Matthew Rea

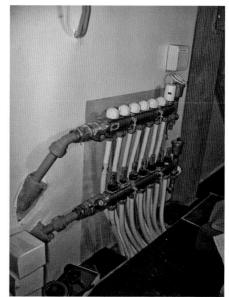

and underfloor heating combined cost around £16,000. This was double a quote they had for £8,000 to install an equivalent oil system. They do not have a meter on the barn, so it is very difficult to separate out the costs of running the heat pump.

The reasons they installed the heat pump were:

- There is no mains gas.
- To have something they could 'fit and forget' with minimum user input.
- They wanted something that could tick-over to provide background heat when the barn wasn't occupied or when they were away.
- They liked the idea that it would be cheap to run after initial outlay.

- They were doing work on the barn and landscaping the garden anyway so the extra upheaval wasn't an issue.
- The cost wasn't all that significant compared to the overall costs of the conversion.
- They liked the idea of installing 'cutting edge' technology with no visual impact.
- They had considered installing PV to provide some electricity but there were no real financial incentives at the time and they were concerned about the visual impact.

However, the family are very happy with the system and the costs for lighting, appliances and heat pump come to an average of around £40 a month. The only small grumble was that they found the user-interface side of it (setting the controls etc.) very complicated. The manifold is under the stairs. The shed outside houses the heatpump and hotwater tank.

Case Study 8: Domestic Ground Source Heap Pump System, Yorkshire, UK

An ICS (www.icsheatpumps.co.uk) barn refurbishment ground source heat pump (GSHP) heating system was installed in Yorkshire, UK, incorporating a heat pump, a buffer tank and a horizontal ground collector. The specified heating system had to match the prestige of the total refurbishment and new build. Being a self-build project the builder also requested that the heating system was from a reputable supplier using reliable high-specification equipment. With this in mind, the heating system specified comprised of a De-Longhi Climaveneta Z1T24 Cube GSHP, a 100 litre buffer tank, polyethylene (PE) pipe horizontal loop ground collector and installation kit to provide all the heating and hot water requirements of the property. Within the building, a high-specification plant room was also incorporated to house the heat pump, completing a textbook installation which really emphasises the low energy and low carbon nature of this type of equipment.

This project exceeded all client expectations and was eventually featured on a well-known TV programme covering high specification renovation projects. The result was a modern alternative to traditional heating with all the benefits of lower energy costs, increased efficiency and lower carbon emissions while being installed within a very traditional-looking building.

The large availability of surrounding land made it ideal for a horizontal GSHP solution drawing low-grade energy via the PE pipe collector and converting it to high-grade heat for the building's hot-water and heating requirements. Factors affecting the design, output and economics of a GSHP system are those that affect performance in a particular locality, especially climate and ground characteristics. The size of the ground loop and the type of ground system utilised are important design factors.

Photovoltaics

Case Studies 9, 10 and 11: Domestic PV Costs and Paybacks

European example without preferential feed-in tariff or net metering

A 3 kWp photovoltaics (PV) system in the south of Europe, which has a similar climate to California, generates 4,500 kWh per year. The cost of installing the system is around €18,000 ($25,000). The system's owner pays €0.15 per kWh to buy electricity from the grid and the electricity company will pay €0.05 per kWh to buy electricity back. The owner uses 30 per cent of the PV generated electricity directly, and exports the remaining 70 per cent.

> Income from exports: 1,500 × 70 per cent × €0.05 = €157.50 per year
> Saving on electricity bills: 4,500 × 30 per cent × €0.15 = €202.50 per year
> Total income and savings: €157.50 + €202.50 = €360
> Simple payback: €18,000 / €360 = 50 years

Clearly this system is not economically viable, even without budgeting for inverter replacement or cost of finance, as the simple payback exceeds the expected panel life (usually 25 years).

UK example with preferential feed-in tariff

A 3 kWp PV system in the UK generates 2,700 kWh per year. The cost of installing the system is £15,000 (around US$25,000). The system's owner pays £0.12 per kWh to buy electricity from the grid and the electricity company will pay £0.03 per kWh to buy electricity back. The owner uses 30 per cent of the PV-generated electricity directly, and exports the remaining 70 per cent. In the best of circumstances (i.e. those who signed up before April 2012), the system owner receives a generation tariff (feed-in tariff) of just over £0.43 for every kWh generated, and this price is guaranteed for 25 years.

> Income from exports: 2,700 × 70 per cent × £0.03 = £56.70 per year
> Saving on electricity bills: 2,700 × 30 per cent × £0.12 = £97.20 per year
> Income from generation tariff: 2,700 × £0.43 = £1161 per year
> Total income and savings: £56.70 + £97.20 + £1161 = £1314.90 per year
> Simple payback: £15,000 / £1,125 = 11.5 years

With the additional support this is clearly an attractive proposition for a householder looking to invest their own savings. Given the long PV module life, and the guaranteed future tariffs, it is also possible that this scheme could be financed through borrowing if the cost of finance is attractive.

All of this is despite the fact that this system does not perform as well as the previous system in terms of annual output, capital cost and the going rate for grid electricity. Feed-in tariffs are designed to be just high enough to attract installations with private finance, at the current cost for installations, so don't expect to make a financial killing. But in a country with a decent feed-in tariff you should be able to install a sensible system, pay off the capital, make a small profit and do your bit towards reducing CO_2 emissions to boot. And you will be going some way towards protecting yourself from future energy price increases too.

USA example with net metering and tax credit (or rebate) on capital costs

A 3 kWp PV system in Kansas, US, is installed with net metering. A bi-directional meter measures how much energy has been exported to and imported from the grid. Any electricity that is generated on-site but not used by the customer is credited to their next bill at retail rate. At the end of the billing year any unused credit is granted to the utility company. In the summer the household generates more electricity than it uses, but the credit from this sees them through the less sunny winter.

The cost of installing the system is US$25,000 but after taking into consideration Federal Tax Credits (30 per cent of the gross cost) the net cost is around $17,500. Over the year the system generates 4,200 kWh of electricity and anything that is not used on site is exported to the grid and is credited to their bill.

> Saving on electricity bills: $0.10 × 4,200 = $420 per year.
> In terms of simple savings, the payback is: $17,700 / $420 = 42 years

Even though the system again performs better than the system installed in the UK the value of the electricity generated is only $0.10 per kWh so the system is still not financially viable.

Had the same system been installed in California, USA, the householder might have been eligible for a Federal Tax Credit of 30 per cent (equivalent to around $7,500) and also a grant of around $7,000 from the local utility (depending on the area and utility company) bringing the net cost down to about $10,500. With a bit more sunshine than Kansas the system would generate around 4,500kWh per year and with net metering would save $450 per year on electricity bills. This would reduce the payback, in simple terms, to 22 years.

There are numerous resources on the web to work out electricity generation potential and costs from PV. See Annex 2 for the full range of contacts and details, and this useful independent website: www.solar-estimate.org. The PV industry would like to see grid parity, where the cost of PV-generated electricity is the same as that bought from the grid by the end user. There are, however, a lot of factors involved: system cost, solar radiation in the area and utility prices in the district. Germany hopes to reach grid parity for PV by the end of 2016, but is aided in this by high electricity prices and a competitive industry.

Case Study 12: PV System on Business Premises, UK

When UK company SWEA moved to its new leased premises in early 2009, there was a strong desire for the company to install its own renewable energy system on or near the building. A quick assessment showed wind and PV to be the most

Figure 9.13a Completed roof-mounted PV installation at SWEA offices, UK
Source: SWEA, www.swea.co.uk

Figure 9.13b Installing hybrid PV module, lifting module onto SWEA office roof
Source: SWEA, www.swea.co.uk

Figure 9.13c Fronius inverters used by SWEA to convert PV output into grid-compatible electricity
Source: SWEA, www.swea.co.uk

relevant technologies. The former was ruled out due to potential planning and land-ownership complications. Local installers were engaged to design the most appropriate type of PV system for the building.

The front of their new building has a variety of roofs facing south-east and south-west, all within 45° of due south, which is generally deemed the viable range for this region of the northern hemisphere. A split system, comprising equal-sized frame-mounted PV arrays on both of the upper roofs, was designed to give excellent all-round performance. Utilising Sanyo HIT hybrid panels – based on monocrystalline cells sandwiched between layers of amorphous silicon – the system improves solar capture under low-light conditions since the dual arrays have different output profiles through each solar day. Fronius IG30 inverters convert DC electricity to grid-synchronised AC electricity, and use maximum power point (MPP) trackers to continually optimise the DC voltage of the arrays for best efficiency.

Each array comprises 14 panels, and is rated at 2.94 kW peak output, giving a total system capacity of around 5.9 kWp. The system managed to generate a total of 5,587 kWh of carbon-free electricity in its first year, resulting in an impressive specific annual yield of 950 kWh/kWp. Nearly 75 per cent of total PV

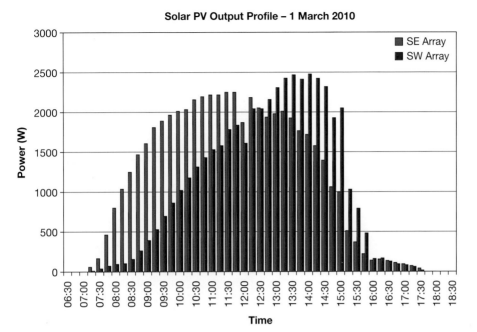

Figure 9.14 Chart showing solar power output profile for the first day in March 2010 at SWEA offices, UK. Note the additional generation time gained by installing two solar module arrays – one on the southeast facing roof and the other facing southwest.

Source: SWEA, www.swea.co.uk

generation occurs during the six months from April to September inclusive. The green power is fed directly into the building's electrical distribution unit, and for much of the time will be directly offsetting the power consumption of SWEA's office lighting and computers, meeting around 20 per cent of their annual consumption. At times such as summer weekends, clean electricity will be spilled into the local grid. Around 25 per cent of the PV system's output is predicted to be exported in this way.

The PV system also benefits from a roof-mounted weather station, which records sunlight hours, solar radiation intensity, ambient temperature, panel temperature and wind speed. All of this information is fed into a data-logger, along with inverter output data, and can be analysed to assess the impact of different weather conditions on performance. The cost of the fully installed system was around £38,000 (around US$60,000) and was funded predominantly by grants.

The project encountered few barriers. Planning permission was required, but gaining this was straightforward. SWEA leases its premises, and so needed to secure permission from the landlord for the required building modifications. These included minor roof strengthening (completed within a single day) that many buildings do not require, tile removal and the addition of wiring conduits. The system yielded over 5,500 kWh in the first full year of operation. The equivalent grid-supplied cost of this would be over £20,000 in the UK. Revenue from the preferential feed-in tariff (FiT) will provide ongoing income from exported electricity and the system has the potential to offset significant grid-electricity usage, reducing power bills further. PV can also play an important role as a distributed energy technology for retrofitting in dense urban areas.

Domestic Grid-connected PV System, Dyffryn Cottage, Wales, UK

Figure 9.15 PV arrays can be seen on two roofs at Dyffryn Cottage's domestic installation

Source: Dilwyn Jenkins

The owners of a small cottage in a village that has good energy-efficiency measures implemented decided to install a 2.15 kWp rooftop PV installation (10 Kyocera 215 polycrystalline modules) with a view to generating clean electricity for use at home and selling any surplus to the grid. The overall installation costs were around £8,000, but the householders have been able to take advantage of the UK's FiTs scheme which offers £0.43p per kWh generated. The cost of electricity from the grid to the household is £0.14p from their suppliers, Good Energy. At the time of writing, the annual electricity bills had been reduced to below £200 a year and the income from the sale of solar PV electricity was around £800 a year.

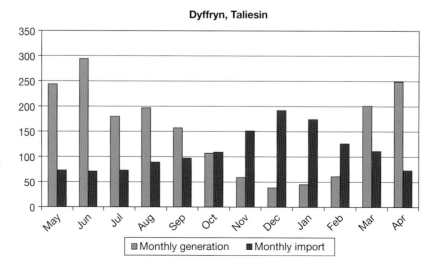

Figure 9.16 Chart showing monthly PV output (generation) against monthly import of grid electricity. There is a marked increase in PV output and decrease in grid electricity used during the summer months

Source: Andrew Rowbottom

Wind Power

Community Wind Power Case Study, Hockerton, UK

After success with an Iskra wind turbine (with 5.4 m blade diameter), the Hockerton Housing Project in the UK joined in with a wider group of villagers from the neighbourhood, with a series of meetings starting in 2006 to look at ways of reducing Hockerton's carbon footprint. As the first stage in a broader plan (which includes energy conservation and demand-side management) it was decided to invest in a community owned 225 kW Vestas wind turbine. A secondhand but fully working machine was procured and installed just four years later. It had been selected to produce roughly as much electricity as the village consumes. Prior to development, studies were made into, and results presented on, a variety of outputs and impacts, covering issues like: power and noise generation, shadow flicker, civil aviation and Ministry of Defence requirements, electromagnetic disturbance, landscape impact, net environmental gain plus impact on wildlife and local employment.

The size of the turbine was calculated to generate enough to cover the electricity consumption by the parish:

- Parish electricity consumption is 275 MWh per year; so
- Turbine production would need to be around 330 MWh per year; which
- Offers a CO_2 equivalent saving of 176 tonnes a year.

According to the developers, the main lesson learned was that the most difficult part of pulling everything together was not at the nuts and bolts end, but rather on the legal and financial side. By the end of December 2010, the turbine had already generated over 240,000 KWh of electricity, an emissions saving of 136 tonnes of CO_2.

It is a community turbine owned by 77 investors, over half of whom are local and who donated an average of around £3,500 each to buy the selected secondhand Vestas wind machine. Led by a group of Hockerton residents, a community company – Sustainable Hockerton (www.sustainablehockerton.org) – was created in 2009 to manage the project with an aim to both help the environment and get a healthy annual return on the money invested in the turbine. The return on the first year was expected to be around 5 per cent, significantly better than the investors could have fund in any bank at that time. There is now a waiting list of people wanting to buy into the scheme.

The energy generated is worth approximately £54,000. As well as paying out to investors this will be used to pay for running costs, with some saved for repairs. Any monies remaining will be given back to the village to support sustainable ventures. Hockerton is also actively encouraging other community groups to consider setting up their own wind turbines and runs a range of courses. Hockerton Housing Project runs a workshop on 'harvesting wind for communities and farmers'.

Figure 9.17 Hockerton Community's 225 kW Vestas wind turbine installed for local investment in a UK parish

Source: Hockerton Housing Project (www.hockertonhousingproject.org)

Hydropower

Community Hydropower, Brecon Beacons, Wales, UK

Hydropower is a well-proven technology, relying on a non-polluting, renewable and local resource, which can also work well with water-supply and irrigation projects. Small-scale, micro hydro offers one of the most environmentally benign energy generation systems because it doesn't interfere significantly with watercourses. However, opportunities for hydropower are often overlooked or abandoned because they are too small scale to be economically viable for large companies. Such opportunities can be ideal as community-owned operations. Talybont Energy, a community organisation based in the Brecon Beacons National Park (Wales, UK), were fortunate to have an abandoned hydro system near to their village that invited attention.

Talybont Energy (www.talybontenergy.co.uk) was established as a not-for-profit community enterprise and company limited by guarantee in order to manage the hydropower project and the income generated on behalf of the community. Since its inception in 2001, and the completion of the hydropower system in 2006, the group has worked to promote, inform, encourage and enable others, locally and nationally, to reduce their environmental impact. In that year, a group of interested individuals from Talybont-on-Usk took a walk along their valley to look at a decommissioned turbine house at the bottom of Talybont Reservoir. Following an official feasibility study showing the project's potential it took two years of ongoing commitment and voluntary time to obtain lease agreements and permissions (i.e. access, planning and abstraction/containment licences).

Figure 9.18 Turning on the valve between the penstock piping and the turbine in the power house at Talybont Energy community hydro-turbine

Source: Talybont Energy

Five years later, and with an exceptional 100 per cent grant funding, the now officially formed group celebrated the opening of an operational, community owned, 36 kW cross-flow hydro-turbine with a head (the distance from the surface of the reservoir and turbine house) of 24 m (80 ft). The theoretical wattage available is 54.2 kW and the turbine flow is about 230 l/s (3,645 gallons per minute) in winter and 115 l/s (1,823 gallons per minute) in summer. The flow is constant, so the turbine output can be accurately predicted. In reality the turbine produces a maximum of 33 kW in winter (60 per cent efficient) and 19 kW in summer. This is produced 24/7 so the turbine produces around 240,000 kWh (220 to 250 MWh) annually. This equates to 60 homes' worth of energy use per year.

With all funding and permissions in place the turbine system was ordered in January 2005. It was installed and in January 2006 a sales contract was formalised with energy supplier Good Energy. In 2009 the turbine generated 241 MWh of electricity, saving about 115 tons of CO_2, and generating over £25,000 which was invested in sustainable projects for the locality, with an emphasis on self-sustaining communal initiatives or facilities, such as electric bikes, PV for the village hall, a zero-carbon car-sharing club with one electric and one 100 per cent bio-diesel car. The community turbine at Talybont (Wales, UK) is also used for environmental education and is a new attraction in the village

This high-head system has proved to be very reliable (high-head systems can have a life-span of 25 years or more). Hydro systems maintenance usually only involves keeping them clean and checking that the amount of flow in the watercourse is acceptable according to the abstraction license. The nine Directors of Talybont Energy (now in its sixth year) are from the local community and volunteer their time to work for the group. They meet once a month for formal committee meetings and employ an admin person four hours a week to help maintain the website, answer e-mails and undertake project work.

10
Annex 1: Finance

Financing Small-scale Renewable Energy Installations

The capital costs of microgeneration installations can be quite high, however there are a number of funding options available to help make renewables economically viable for householders and small businesses. Grants and loans are available from a number of sources including the government, local authorities, banks and energy companies. There are also government tariffs and tax incentives that can help to significantly reduce the payback time on your investment. With so many schemes currently running it is likely that there will be some form of support available in your country or region. An easy way to check is to enter your details into the Energy Saving Trust's Grants and Discounts Database, which will provide you with information on funding in your area that has technology and property-specific requirements (www.energysavingtrust.org.uk/Easy-ways-to-stop-wasting-energy).

It is also well worth contacting your local authority. As part of government initiatives to meet the UK's legally binding target of 15 per cent of energy from renewable sources by 2020, many local authorities are running grant and loan programmes to encourage residents to adopt low carbon, renewable technologies. Funding streams are usually limited and high demand can mean that grant schemes in particular get snapped up very rapidly, so it is a good idea to act fast when applying.

A large proportion of funding is set aside for community projects as opposed to individual home and business instalments. It may be worth exploring the idea of carrying out your installation as part of a community project: this could mean either initiating a new project or joining an existing one. Being involved in a community energy scheme not only opens up many funding streams but can also provide you with a network of advice and support both during the installation process and after.

Perhaps the most important financial incentives for renewable electricity generation are the preferential feed-in tariffs (FiTs) available in many countries. These are covered in the Performance-linked Incentives section below. In the US rebates on the capital costs of installation are also available in many states.

Financial support in the US

The US Department of Energy's (DOE) Energy Efficiency & Renewable Energy (EERE) programme coordinates funding for renewables projects in the US (www1.eere.energy.gov). There are several tax credits, grants and rebates available to individuals wishing to install microgeneration technologies in the US. A national tax-credit scheme is currently running that offers around 30 per cent credit on biomass, heat pumps, solar, wind and fuel cells, and will be active until 2016. The government and some private loan companies are also offering Energy Efficient Mortgages, which can help buyers renovate an inefficient property or build a new efficient property. Such mortgages can be used to fund microgeneration as part of a larger whole-house energy efficiency project.

There are also several state-based schemes operating throughout the country offering incentives for renewable generation. The funding available, and its focus, varies greatly between states, with some prioritising environmental issues more than others. For example, California runs an FiT scheme similar to that in UK, while New York offers grants of up to 50 per cent for renewable microgeneration instalments. The climate also affects the types of funding available, with southern states focusing on solar power while northern states tend to push heating and insulation measures. DSIRE, the Database of State Incentives for Renewables and Efficiency (www.dsireusa.org), can tell you what is available in your area.

The DOE EERE also runs a Tribal Energy Programme that promotes energy efficiency and sustainable development on tribal lands through various financial and technical assistance schemes. Government grants, loans and investment schemes are available for people on tribal lands wishing to develop renewable energy generation projects.

In the US there are sometimes small electricity buy-back credits available for renewable power fed back into the grid. The rate and its availability varies from state to state and year to year, but information is available from your local utility. There are two programmes relevant to small-scale power producers: the National Energy Policy Act (1992) and the Public Utilities Regulatory Policy Act (PURPA).

Financial Support for Domestic Installations, Small Businesses and Communities in the UK

Many of the funding streams available are open to both individuals and small businesses. Community projects, for which much of the funding is set aside, usually involve both households and businesses. When researching funding opportunities your first port of call should be your local authority. Many local authorities are running schemes to encourage residents to adopt energy efficiency measures in their homes. Some companies, such as Affordable Energy, specialise in working with local authorities to deliver council-funded renewable energy schemes and will carry out the research for you.

National Grants and Loans

The Low Carbon Buildings Programme offers grants for installations in the public sector and non-profit organisations. Community Energy Scotland's Community and Renewable Energy Scheme (CARES) has a a loan scheme for community energy projects. The EDF Green Fund (www.edfenergy.com) supports community and non-profit organisations with the costs of renewable energy projects. The E.ON Sustainable Energy Fund has been providing grants to support community microgeneration projects in the UK.

As part of the government's Carbon Emissions Reduction Target (CERT), large energy suppliers are required to promote energy saving measures to their consumers. This means that energy suppliers are obliged to provide grants and discounts to help you install energy saving measures in your home. Many energy companies also offer free loft and cavity wall insulation for vulnerable groups such as those over 70 and those in receipt of particular benefits.

Some companies are offering free solar (PV) deals where solar panels are installed free of charge and free energy is generated for your home on the condition that the company receives the full FiT income. In accepting such a scheme you will be missing out on the potential income that can be generated from microgeneration: the Energy Saving Trust provide consumer guidance for free PV on their website with suggestions on what to ask for and what to expect from the company (details of which can be found at the end of the chapter).

If you are planning an installation that is eligible for government incentive schemes it is relatively easy to show banks that the installation will generate reliable income and savings which can be used to pay the loan instalments. There are also private finance companies such as Renewable Funding (www.renewablefunding.co.uk) that specialise in financing renewables installations. The BIG Lottery Fund provides funding for community environmental projects.

The government is currently running two performance-linked incentive programmes as part of the Clean Energy Cashback scheme – the FiT scheme for renewable electricity and the RHI for renewable heat energy – both of which are available in different capacities to householders and businesses (more details in the section on performance-linked incentives). While much of the funding is open to both households and businesses (often as part of a community) some financial support schemes have be designed specifically to meet the needs of small businesses. For example, Energy Efficiency Financing (www.energyefficiencyfinancing.co.uk) offers leases, loans and other financial services to organisations wishing to reduce their energy use.

The Energy Saving Trust Scotland offers interest-free loans to small businesses to help them install renewable energy generation technologies. The Enhanced Capital Allowance (ECA) is a tax-incentive scheme aimed at encouraging businesses to invest in energy saving technologies: 100 per cent of capital expenditure of the instalment can be deducted against taxable profits in the first year. Although it is aimed at larger industry it can also be relevant to some smaller businesses. The Energy Technology List found on the ECA website can help you find eligible equipment for your installation (www.energysavingtrust.org.uk/scotland).

Financial Support in Canada

Natural Resources Canada's Office of Energy Efficiency (OEE) offers a number of grants and incentives to encourage the uptake of renewable energy technologies.

The ecoENERGY Retrofit programme offers financial support to encourage the implementation of energy-saving projects. There are also numerous complementary regional programmes such as incentive schemes and rebates that have been designed to work with the ecoENERGY scheme to further support residents in their projects (Canada OEE Complimentary Regional Programmes, www.oee.nrcan.gc.ca). Many of these regional schemes are also available for businesses and organisations. The OEE has an interactive map that allows you to easily search for funding opportunities in your province or territory.

Performance-linked Incentives

If a wind system, PV or hydro system is designed to generate a surplus of electricity over and above the requirements of a home or business, then this can in theory be exported and sold into the national grid. A qualified or accredited installer/electrician is needed to connect a wind turbine's output to the mains in order to get a grant under most incentive schemes.

Feed-in Tariffs

Feed-in tariffs (FiTs) are designed to encourage domestic, business and community energy users to generate their own clean electricity. To do this energy suppliers pay small-scale generators for electricity from renewable or low-carbon sources, including wind turbines, solar PV, micro hydro, micro CHP and anaerobic digestion. This scheme guarantees both a minimum payment for all electricity generated and a separate payment for the electricity exported to the grid. On top of this, the property will experience smaller bills due to grid-imported electricity now being offset by the generator's output. If a renewable energy system is specified to produce more than the property consumes, over the year the bills should be positively in favour of the generator.

In the UK, in order to sell electricity into the grid you will need to invest in a special meter (about £120) as well as negotiate a rate with your electricity supplier. Rates vary from supplier to supplier and also between state policies (in the US) and national policies (in Europe), whether government or utility based. If the supply is liberalised and competitive it should be possible to shop around and find a sympathetic supplier. At the heart of FiTS is a special two-way metering concept that allows generators to sell to the grid at a higher rate per kWh than they buy from it. How much a renewable energy system might earn under the scheme depends on the system type and size as well as the available renewable energy resource, although the figures quoted are that, generally over the lifetime of the tariff, a generator might expect to recover the capital investment more than twice over. FiTs are planned for up to a 25 year period, depending on the renewable technology (e.g. wind is for a 20 year period).

In the UK a Microgeneration Certification Scheme (MCS) certifies microgeneration products under 50 kW as well as the approved installers. Systems over 50 kW need to apply directly through the Renewables Obligation Order FiT process (ROO-FiT) process which is outlined on the Ofgem website (www.ofgem.gov.uk). It's a good idea to check with the Energy Saving Trust and/or Ofgem about the scheme, since the UK Government's Department of Energy and Climate

Change (DECC) have reviewed and made changes to the original FiTs proposals. The EST offer an example of how an FiT works in a fairly typical domestic solar electricity system, with an installation size of 2.7 kWp. In money terms, the system could make around £1,170 per year assuming that 50 per cent of the electricity generated is exported.

How It Works

The FiTs scheme essentially pays small-scale renewable electricity generators for the electricity they produce. FiTs are available to all individuals and organisations with an installation capacity (the maximum system output) of less than 5 MW in the UK excluding Northern Ireland (under review). This is plenty for a household or small business (installations with a capacity greater than 5 MW will fall under the Renewables Obligation scheme). The scheme includes solar PV, wind turbines, hydroelectricity, micro combined heat and power (CHP) and anaerobic digestion. If you choose to have multiple systems on a property, each system will be dealt with and metered separately, so there is no maximum capacity for each claimant but only for each installation. Technology ownership is linked to the site, therefore if you move house the FiT payments will not move with you and likewise if you buy a house with an FiT-eligible instalment already in place you will be able to collect the payments. The scheme is funded by the six main energy companies in the UK (known as the 'big six') who are obligated by the government to pay householders and communities a minimum price for each unit of renewable energy that they produce (measured in kWh).

A qualified or accredited installer/electrician is needed to connect a renewable energy system's output to the mains in order to get a grant under most incentive schemes. For safety's sake, grid-connect wind systems will automatically shut down if there's a grid power-cut in the local electricity distribution network. The reason for this is that small wind turbines need batteries or a grid to smooth out fluctuations in the consumption and generation of electrical power.

In the US, there are sometimes small electricity buy-back credits available for wind power fed back into the grid. The rate and its availability varies from state to state and year to year, but information is available from your local utility company.

Australia operates a form of net metering, offering the standard FiT for people producing power for their homes or small businesses using renewable energy systems with a capacity of up to 100 kW. This includes wind, solar, hydro or biomass power. Excess power fed back into the grid is credited at the same retail rate charged for electricity consumed. Australia also operates a Solar Credits mechanism under the expanded Renewable Energy Target (RET) that multiplies the number of Renewable Energy Credits (RECs) able to be created for eligible installations of small generating units. Credits belong to the owner of the renewable energy system, however, to save you the work of trading the RECs yourself, installers sometimes buy these credits from you. The value of the RECs under the current scheme fluctuates over time. The Australian Renewable Energy Regulator publishes a 'Small Generators Owners Guide', which describes how to calculate how many RECs your system could earn, but REC income is linked to the rated generating capacity of the renewable generator.

Payments are usually delivered directly to you from your energy supplier every quarter although this varies between suppliers and instalments. In some cases your installation company may offer to collect the tariffs on your behalf and use them as part payment for installation costs. In the UK, both the generation and export tariff payments are exempt from income tax for households. However, this is not the case for businesses, and income received must be declared as part of taxable revenue, although some businesses may be eligible for Enhanced Capital Allowances.

Joining a FiT scheme should significantly decrease the payback period of your installation. Not only are you saving on your bills by buying less electricity from the grid but you are also receiving an income from your installation. To put the numbers in perspective a typical 2.5 kW, well-sited solar PV system in the UK could lead to savings of up to £140 on your annual energy bill while income from the electricity generation and export tariffs could be as much as £900 per year (see DECC FiT leaflet: www.decc.gov.uk). As it is assumed that many people will take out loans to fund renewable installations the tariff has been designed so that it should provide a monthly income that is significantly higher than your monthly loan repayments.

Calculating Payback

The payback period (from savings and/or income from electricity sold) is a useful way of assessing the viability of an investment and also allows you to compare the financial merits of different technologies and options. It does not take into account all the factors such as non-financial costs and benefits, but is a good way of determining where your money will be spent most effectively.

To carry out a simple estimate of the payback on a proposed installation you need to divide the total costs of the installation by the combined annual savings and income. For example, if your ground source heat pump is going to cost you £12,000 to install and is replacing coal which will give you annual savings of £600, your estimated payback would be 20 years. Calculating the internal rate of return (IRR) for your investment may also be useful and will provide you with a more complex analysis of the estimated return on your installation. However, a simple payback calculation should be sufficient to inform an investment decision. Variables that will affect the payback period include:

- *Energy prices*: it is predicted that rising oil and gas prices, and the shift away from fossil fuel generation to renewable generation, will lead to rising energy prices for consumers. This will increase the saving made by generating your own energy.
- *The system being replaced*: savings will depend on the efficiency and costs of the previous system. Replacing an oil burner with a ground source heat pump will yield much greater savings and a much faster payback period than replacing a gas system with the same technology.
- *The efficiency of the new system*: the efficiency of the system being installed will have a big impact on payback times. A poorly sited PV system simply will not generate enough energy to payback in its lifetime even though the equipment itself has the capacity.

- *The energy efficiency of the property*: combining a renewable installation with energy efficiency improvements can help increase savings. Although the costs of insulation must also be included in the payback calculations, relatively small investments in insulation can have a large impact on energy demand.
- *Incentives*: receiving a tariff such as the FiTs or the RHI can take a significant chunk out of your payback time as you are earning on top of the savings.

The number of variables involved make it difficult to make generalised payback period estimates. A study by the National House Building Confederation (NHBC) on microgeneration arrived at the following estimates. It should be noted that they do not take FiTs or the RHI (see Table 10.1 for more information) into account and so payback would be significantly shorter than shown for eligible installations.

Averages such as these can offer a guideline, but they are very limited as they don't take into account the specifics of your installation, which may be well suited to your site or wholly inappropriate. To get a better idea of the payback period for a particular installation proposal the Energy Saving Trust provides a useful Cashback Calculator tool. It takes into account site, scale, costs, energy use and available incentives, and provides you with estimated savings and an estimated payback period. It is well worth consulting this tool as part of your research so that you can get a rough idea of the kind of savings and payback period you can expect from your instalment.

Some benefits are not included in the payback time but are still of importance when making an investment decision. For example, microgeneration can be considered a home improvement and thus should increase the value of your property if you decide to move. You may also want to have a look at the energy and carbon payback times as these can give an idea of the environmental rewards

Table 10.1 Estimated financial paybacks for each technology (assuming no tax credits, grants, FiTs, or other financial support)

Technology	Estimated payback period (years)
Biomass	N/A fuel costs on a par with oil but capital expenditure for installation was higher making no savings. Access to free or cheap fuel such as free wood would alter this.
Solar PV	Savings made but not able to payback during its lifetime.
Solar thermal	8–20 depending on site.
Wind	15 for well-sited 50% grant funded 2.5 kWh turbine.
GSHP	8–15
ASHP	8–15
Hydroelectric	15 for a 100 kW installation with income from ROCs (Renewable Obligation Certificate).
Micro CHP	3–5

Source: NHBC Foundation (2008) 'A review of Microgeneration and Renewable Energy Technologies' http://www.nhbcfoundation.org

Table 10.2 Energy and carbon paybacks by technology

Technology	Energy payback time (years)	Carbon payback time (years)
Micro-wind	1.6–2.5	1.1–1.7
Solar PV	5.0	6.0–8.0
Solar water heating	2.0–3.8	2.4–4.4
Ground source heat pump	2.1	6.0
Biomass boiler	N/A	0.6–1.1

Source: EST (2007) 'Generating the Future: An Analysis of Policy Interventions to Achieve Widespread Microgeneration Penetration' available at: http://www.energysavingtrust.org.uk

of different installations in addition to the financial rewards. Table 10.2 illustrates the lifecycle payback periods (the time taken for carbon and energy savings made to equal the amounts involved in the production, use and disposal processes) of several renewable technologies.

When thinking about the payback of your installation it is also sensible to consider the non-financial benefits that can be gained from microgeneration. Producing energy from your land and knowing that you are partly energy self-sufficient can be very satisfying. By installing microgeneration technology you are taking an active role in reducing your carbon footprint and encouraging others to do the same, both of which are crucial to tackling climate change. Renewable installations can be costly, but for many the benefits reaped from investing in a healthier and more sustainable world go far beyond the figures.

The Renewable Heat Incentive (RHI)

To make solar water-heating even more attractive, there are an increasing amount of government incentives for the take up of solar thermal. In the UK for instance, after considerable pressure on the part of environmental groups, it was finally announced in March 2011 that the UK would be pioneering the world's first Renewable Heat Initiative (RHI), which will pay householders for the renewable heat produced by their solar water system as long as it's approved under the MCS scheme. Heat accounts for 47 per cent of the UK's energy consumption so it is essential that renewable heat sources are adopted if the UK is to meet its legally binding target of 15 per cent renewable energy by 2020 (DECC RHI, http://www.decc.gov.uk). The aim of the RHI is to encourage householders and businesses to switch from fossil fuel heating to renewable sources such as heat recovery (heat pumps), sustainable biomass and solar thermal.

The incentive is designed to bridge the gap between the costs of conventional fossil fuel heating and renewables, making the adoption of renewable heat affordable. It is designed to help householders by subsidising the upfront costs of installations with a one-off payment. In order to be eligible it is essential that your home meets the energy efficiency criteria of the scheme. This will be determined based on the Energy Performance Certificate (EPC) of your home, which comes in the form of a letter grading from A to G (A being a very energy efficient property and G being a very inefficient property). You can commission an EPC assessment yourself for a fee of around £60, and the EPC register will help you find an accredited assessor (see Annex 2 for details). Criteria aside, if you are considering a renewable heat installation it is essential that your home is energy efficient, otherwise the financial and environmental benefits of any installation will be undermined.

During the first phase of the RHI, businesses and other non-domestic installations such as schools, not-for-profit organisations and community groups will be eligible for a long-term (20 year) tariff. This means that they can receive a payment for every unit of heat (measured in kilowatt thermal or kWth) that is generated. DECC defines a domestic installation as serving one private residential dwelling only. This means that many small businesses, including houses that have been converted for non-residential use (as in the case of a bed and breakfast), are included in the first phase of the tariff scheme, as are public-sector buildings such as schools, and community heating projects where several properties are served by one installation.

In the second phase (2012), a parallel initiative, the Green Deal for Homes, offers help to improve insulation and energy efficiency across the UK. The Green Deal should bring with it opportunities to compound benefits by taking a 'whole-house' approach and combining government-funded energy efficiency improvements with an RHI eligible renewable heat instalment. At the time of writing, further details of the second phase are due to be announced shortly, so it is worth checking the DECC and EST's websites regularly for updates.

To receive both the RHI tariff and the Renewable Heat Premium Payment fund, you will need to apply to Ofgem. Thus can be done online, by phone or by post once the scheme has started. You will be required to show that the heat produced will be useful and not wasted and that your property has a high level of energy efficiency. Installations up to 45 kWth (kilowatt thermal or kilowatt heat) will be required to show that both the equipment and the installer are MSC certified (systems larger than 45 kWth will be assessed separately by Ofgem). Applicants will also be required to agree to some ongoing obligations which include keeping up with maintenance, allowing inspections by Ofgem and providing relevant information on their system to make sure that eligibility criteria continue to be met throughout the lifetime of the tariff.

Eligible technologies for both phases will include MSC certified biomass boilers, solar thermal and ground and water source heat pumps (air source heat pumps have initially been excluded from the RHI tariff pending further research but are eligible for Premium Payment support). Cooling provided by heat pumps is not included in the scheme, although heat used for cooling via absorption chillers in industrial processes is eligible for RHI support. If you are considering a heat pump it is important to note that heat pumps will need to have a COP of 2.9 or higher in order to qualify for the tariff (the manufacturer will be able to provide you with this information). The amount of heat generated will be measured using heat meters. As the scheme progresses it is likely that tariffs will be reduced each year for new entrants (known as degression). However, tariffs for instalments already on the scheme will remain unchanged for the duration of the scheme.

Benefits

Savings made will depend greatly on the type of heating system that is being replaced. The RHI will be particularly beneficial for those who are not on the gas grid, as such properties will tend to use high-cost and high-carbon fuels such as oil and coal and will thus see the greatest savings. Income from the RHI should significantly reduce the payback time of your instalment, especially if the capital costs have been subsidised by the Premium Payment fund.

Funding Options

US

- *Database of State Incentives for Renewables and Efficiency (DSIRE)*: state-by-state listings of funding for renewables, http://www.dsireusa.org/
- *DOE EERE*: funding opportunities for households and small businesses, www1.eere.energy.gov/financing/consumers.html
- *DOE EERE Tribal Energy Programme*: funding opportunities on tribal lands, http://apps1.eere.energy.gov/tribalenergy/

UK

- *Affordable Energy*: working with local authorities, Affordable Energy delivers council funded renewable energy and insulation grant-schemes to households across the UK. They also offer a free no obligation survey of your home to assess which renewable energy generation technologies may be most suitable, www.affordableenergy.co.uk/generate-energy.php
- *Big Lottery Fund*: www.biglotteryfund.org.uk/funding-uk
- *Carbon Emissions Reduction Target (CERT)*: www.decc.gov.uk/en/content/cms/what_we_do/consumers/saving_energy/cert/cert.aspx
- *The Centre for Sustainable Energy*: information on renewable energy grants in the South West, www.cse.org.uk/pages/energy-advice/grants/renewable-energy-grants/
- *Community Energy Scotland*: www.communityenergyscotland.org.uk/grant-funding.asp
- *Department of Energy and Climate Change (DECC)*: for policy information and details of tariff banding, www.decc.gov.uk/en/content/cms/meeting_energy/Renewable_ener/feedin_tariff/feedin_tariff.aspx
- *Energy Efficiency Financing*: a partnership between the Carbon Trust Implementation Services and Siemens Financial Services. The scheme offers leases, loans and other financial services to organisations wishing to reduce their energy use, www.energyefficiencyfinancing.co.uk/customers/pages/default.aspx
- *Energy Saving Trust*: free PV scheme consumer guidance, www.energysavingtrust.org.uk/Generate-your-own-energy/Solar-electricity/Consumer-guidance-on-free-solar-PV-offers
- *Energy Saving Trust's Grants and Discounts Database*: www.energysavingtrust.org.uk/cym/Easy-ways-to-stop-wasting-energy/Energy-saving-grants-and-offers/Grants-and-Discounts-Database

- *Energy Saving Trust Scotland*: www.energysavingtrust.org.uk/scotland/Scotland-Welcome-page/Business-and-Public-Sector-in-Scotland/Grants-loans-awards/Small-business-loans
- *Enhanced Capital Allowance*: provide tax credits for businesses on the purchase of renewable technologies, www.eca.gov.uk/etl
- *Feed-in-tariffs (FiTs)*: clear information on the scheme, www.fitariffs.co.uk/
- *Ofgem*: for up-to-date information on tariffs and administration, eligability, registration, www.ofgem.gov.uk/Sustainability/Environment/fits/Pages/fits.aspx
- *The Renewable Heat Incentive (RHI)*: www.decc.gov.uk/en/content/cms/what_we_do/uk_supply/energy_mix/renewable/policy/incentive/incentive.aspx
- *Carbon Emissions Reduction Target (CERT)*: www.decc.gov.uk/en/content/cms/what_we_do/consumers/saving_energy/cert/cert.aspx
- *Energy Saving Trust's Grants and Discounts Database*: www.energysavingtrust.org.uk/cym/Easy-ways-to-stop-wasting-energy/Energy-saving-grants-and-offers/Grants-and-Discounts-Database
- *HEEP energy efficiency funding scheme*: www.heepgrant.org/sustainable_energy.html
- *Herefordshire Council energy efficiency grants*: www.herefordshire.gov.uk/environment/land/30902.asp

Canada

- *Office of Energy Efficiency (OEE)*: listings by territory or province of Complimentary Regional Programmes including incentives and grants for renewable energy generation, http://oee.nrcan.gc.ca/residential/personal/retrofit-homes/provincial-municipal.cfm?attr=4
- *OEE ecoENERGY Retrofit scheme*: details of the nation-wide funding scheme for energy efficiency retrofits including renewables installations, http://oee.nrcan.gc.ca/corporate/incentives.cfm?attr=0

11

Annex 2: Resources

Annex 2 is dedicated to signposting access to the main sources of information used in researching this book. At the same time, this section identifies further resources on the renewable energy technologies covered in this book.

General information

North America

Database of State Incentives for Renewables and Efficiency (DSIRE), US: for state-by-state listings of funding for renewables, www.dsireusa.org

Office of Energy Efficiency, Canada: website provides information and advice including details of the nation-wide funding scheme for energy efficiency retrofits and renewable energy installations, http://oee.nrcan.gc.ca

US Federal Energy Regulatory Commission (FERC): www.ferc.gov

United States Department of Energy, Energy's Efficiency and Renewable Energy Office (USDOE EERE): www.eere.energy.gov/

Europe

Carbon Trust, UK: for carbon saving information and services aimed at businesses and the public sector, www.carbontrust.co.uk

Centre for Alternative Technology (CAT), UK: for information and courses on renewables generation. CAT also runs a year round visitors centre, www.cat.org.uk

Centre for Sustainable Energy (CSE), UK: a research and advice centre for renewable energy. Particularly useful for those in the South West of England but has lots of relevant information for people across the UK, www.cse.org.uk

Department of Energy and Climate Change (DECC), UK: www.decc.gov.uk

Energy Saving Trust, UK: information and advice on energy efficiency and renewable energy, including an energy tool selector which can help you determine which technology is best suited to your home, www.est.org.uk/

Government Greener Homes Planning Portal, UK: this is a useful tool for finding out what is permitted by regulations and what is required of you when installing renewable energy technologies, www.planningportal.gov.uk/planning/greener homes

Microgeneration Certification Scheme, UK: search the Microgeneration Certification Scheme database for MCS certified technologies and installers, www.microgenerationcertification.org/consumer

Renewable Energy Association, UK: represents the renewable energy industry in the UK, www.r-e-a.net/

The Renewable Energy Centre, UK: a good information resource, www.the renewableenergycentre.co.uk

Courses

Brighton Permaculture Trust, UK: run several events on sustainable living and renewables throughout the year in the Brighton area, www.brightonpermaculture.org.uk

Centre for Alternative Technology, UK: CAT run a number of short courses as well as a post-graduate course on renewable energy systems and installation, www.cat.org.uk

Centre for Sustainable Energy, UK: CSE regularly run workshops in schools and communities on sustainable energy, www.cse.org.uk

Eco Building Magazine, UK: offers up-to-date listings of events and courses related to green building in the UK, http://ecobuildingmag.com

Homepower Magazine, USA: lists renewable energy courses taking place around the US and Canada, www.homepower.com/resources/events

Low Carbon Trust, UK: LCT run a number of events and seminars throughout the year on renewable energy generation and energy saving, www.lowcarbon.co.uk/events

Mid West Renewable Energy Association, USA: offers a range of courses and workshops on renewable energy technologies, www.midwestrenew.org

National Self Build and Renovation Centre, UK: offer courses on self-build and renovation including renewable heat and energy technology installation as well as listings of eco-build events, www.buildstore.co.uk/mykindofhome/seminars-events.html

Renewables Academy, Germany: offers training in the fundamentals of renewable energy and energy efficiency, www.renac.de

Solar Energy International, USA: renewable energy courses and workshops in the US, including on-line courses, www.solarenergy.org

Sustainable Building Association, UK: offers listings of courses and events related to sustainable home improvement and development, www.aecb.net/index.php

Sustainable Homes, UK: offers many courses and events although these are geared more towards industry professionals than consumers, www.sustainablehomes.co.uk/index.aspx

Books and Publications

General

EERE: an overview of US green power markets: www.eere.energy.gov/greenpower/markets.

Homeowner's Guide to Renewable Energy, Dan Chiras, 2006, The New Society, British Colombia.

Home Power Magazine: US-based magazine and website with lots of information on small-scale renewable energy, www.homepower.com

Off the Grid: Managing Independent Renewable Electricity Systems, Duncan Kerridge with Dave Linsley Hood, Paul Allen and Bob Todd, 2006, CAT Publications.

Sustainable Home Refurbishment, David Thorpe, 2010, Earthscan Expert Series, Earthscan, London.

Wood Pellets

The Pellet Handbook, Ingwald Obernberger and Gerold Thek, 2010, Earthscan, London.

Wood Pellet Heating Systems, Dilwyn Jenkins, 2010, Earthscan Expert Series, Earthscan, London.

Solar Thermal

Planning and Installing Solar Thermal Systems: A Guide for Installers, Architects and Engineers, German Solar Energy Society, 2009, Earthscan, London.

Solar Domestic Water Heating, Chris Laughton, 2010, Earthscan Expert Series, Earthscan, London.

Solar Technology, David Thorpe, 2011, Earthscan Expert Series, Earthscan, London.

Solar Water Heating: A Comprehensive Guide to Solar Water and Space Heating Systems, Ramlow and Nusz, New Society, 2010, Mother Earth News Wiser Living Series, BC: Canada.

Heat Pumps

A Buyer's Guide to Heat Pumps, 2011, Energy Saving Trust, London, offers advice on choosing technologies and suppliers, www.energysavingtrust.org.uk/Resources/Publications/Renewables/A-buyer-s-guide-to-heat-pumps

Domestic Ground Source Heat Pumps: Design and Installation of Closed Loop Systems, Ground Source Heat Pump Association, 2004, www.gshp.org.uk/documents/CE82-DomesticGroundSourceHeatPumps.pdf

Geothermal Heat Pumps: A Guide for Planning and Installing, Karl Ochsner, with introduction by Robin Curtis, 2007, Earthscan, London.

Geothermal HVAC: Green Heating and Cooling, Jay Egg and Brian Clark Howard, 2011, McGraw-Hill, Columbus, OH, USA: comprehensive and detailed book for those wishing to get an in-depth understanding of geothermal heating and cooling technologies.

Heat Pumps for the Home, John Cantor and Gavin Harper, 2011, Crowood Press, Wiltshire, UK.

Heat Pump Technology, Billy Langley, 2002, Prentice Hall Inc., New Jersey, USA.

Photovoltaics

The Complete Idiot's Guide to Solar Power for Your Home, Dan Ramsey, 2003, Apha Books, New York.

The Easy Guide to Solar Electric Part II, Installation Manual, Pieper Adi, 2003, ADi Solar, Santa Fe, USA.

Grid-connected Solar Electric Systems, Geoff Stapleton and Susan Neill, 2011, Earthscan Expert Series, Earthscan, London.

The Solar Electric House: Energy for the Environmentally-Responsive, Energy-Independent Home, Steven J. Strong, 1994, Sustainability Press, USA.

Solar Technology, David Thorpe, 2011, Earthscan Expert Series, Earthscan, London.

Stand-alone Solar Electric Systems, Mark Hawkins, 2010, Earthscan Expert Series, Earthscan, London.

Wind Power

'Energy Saving Trust Field Trial of Small Wind Turbines': www.energysavingtrust.org.uk/Generate-your-own-energy/Energy-Saving-Trust-field-trial-of-domestic-wind-turbines

'Results of monitoring of UK rooftop wind generators, 2009', www.warwickwindtrials.org.uk

'Reviews over 100 wind turbines', www.bettergeneration.com/wind-turbine-reviews.html

'Urban Wind Turbines: Guidelines for Small Wind Turbines in the Built Environment', an Intelligent Energy for Europe Report, 2007, European Commission, Brussels, www.urban-wind.org/pdf/SMALL_WIND_TURBINES_GUIDE_final.pdf

Wind Power Workshop: Building Your Own Turbines, Hugh Piggott, 1997, 2000 & 2011, Centre for Alternative Technology, Machynlleth, Wales.

A Wind Turbine Recipe Book, Hugh Piggott, www.scoraigwind.com, 2009.

Hydropower

British Hydropower Association Guide to Hydro Power, www.british-hydro.org/mini-hydro/download.pdf

Going with the Flow: Small Scale Water Power, Bill Langley and Dan Curtis, 2004, CAT Publications, Machynlleth, Wales.

Hydropower Engineering Handbook, John S. Gulliver & Roger Arndt, 1991, McGraw-Hill, Maidenhead, UK.

Micro-Hydro Design Manual: A Guide to Small-Scale Water Power Schemes, A. Harvey, et al., 1993, Intermediate Technology Development Group Publications, London.

Micro Hydro Power Sourcebook, A. Inversin, National Rural Electric Cooperative Association (NRECA), 1986, Arlington, VI, USA.

'Microhydropower Systems: A Buyers Guide', Natural Resources Canada, 2004, Ottawa, ON, Canada.

Wood Energy

Organisations

North America

Environment Canada: on air quality and emissions, www.ec.gc.ca/residentiel-residential/default.asp?lang=En&n=E9FE1750-1

Natural Resources Canada's 'EcoENERGY Retrofit' Program: offers advice on fitting new wood energy appliances, http://oee.nrcan.gc.ca/residential/personal/retrofit-homes/retrofit-qualify-grant.cfm

US Department of Energy (DOE): www.energysavers.gov/your_home/space_heating_cooling/index.cfm/mytopic=12570

Wood Pellet Association, Canada: part of the Renewable Energy Association, http://www.r-e-a.net

Europe

Biomass Energy centre: www.biomassenergycentre.org.uk

European Biomass Association: http://www.aebiom.org/

Pellet Fuels Institute, Europe: http://pelletheat.org/

Other Resources

UK Smoke Control Areas: in the UK during the 1950s and 1960s city smog was a serious health hazard in the UK. To combat this, the Clean Air Acts of 1956 and 1968 were introduced. These acts evolved into the 1992 Clean Air Act. The legislation imposed Smoke Control Areas within which only authorised 'smokeless' fuels could be used in heating appliances: typically these include natural gas, electricity and anthracite. Within a smoke-control zone, wood can only be burnt on Exempted Appliances pursurant to Smoke Control (Exempted Fireplaces) Orders. It is illegal to burn wood in any other device and illegal to deliver wood (for burning) to properties within the area.

Operated by local authorities, the Smoke Control areas presently cover most of England. Cornwall, Northern Scotland and Wales are mostly free of smoke control zones, but it's important to check the status of your location before selecting which wood heating-appliance you want to invest in. The main reason to check this early on in the process is because there is a growing list of Exempted Heating Appliances.

Exempt appliances are appliances (ovens, wood burners and stoves), which have been exempted by Statutory Instruments (Orders) under the Clean Air Act 1993 or Clean Air (Northern Ireland) Order 1981. These have passed tests to confirm that they are capable of burning an unauthorised or inherently smoky solid fuel without emitting smoke. New additions to this list in 2010 have included many wood log stoves, some log boilers, a few pellet stoves and several pellet boilers.

Solar Thermal

Organisations

The **US DOE** provides information on solar water heating on its website: www.energysavers.gov/your_home/water_heating/index.cfm/mytopic=12760

Flasolar: have an online calculator for sizing solar absorber systems for swimming pools: www.flasolar.com/php/pool_panels_entry.php

Florida Solar Research Centre: has excellent information on solar pool heating, including sizing and economics, installation and system ratings: www.fsec.ucf.edu/en/consumer/solar_hot_water/pools/index.htm

The European Solar Thermal Industry federation: http://www.estif.org/

Solar Trade Association, UK has a useful website and a free downloadable Solar Handbook (they cover solar thermal and PV): www.solar-trade.org.uk

In the UK, there are what is known as **Renewable Heat Incentives** which provide small grants towards solar thermal systems and a generation tariff which pays a sum for every kilowatt hour of heat energy produced. The following link provides more information about this: www.decc.gov.uk/en/content/cms/meeting_energy/Renewable_ener/incentive/incentive.aspx

Other Resources

DOE's Energy Savers on solar ('active') space heating: www.energysavers.gov
Homeowner's guide to passive solar, available free online: www.builditsolar.com/Projects/SolarHomes/PasSolEnergyBk/PSEbook.htm
RETScreen Software Solar Water Heating Model has worldwide weather data to evaluate the energy production and savings, costs, emission reductions, financial viability and risk for solar water heating projects using Microsoft Excel: www.retscreen.net/ang/g_solarw.php
Solar evacuated tubes for swimming pool applications in the UK: www.solaruk.net/lazer2_solar_thermal_collectors.asp
Solar Radiation Data Manual for Flat-Plate and Concentrating Collectors in the US, Pacific Islands & Puerto Rico as data in pdf format: http://rredc.nrel.gov/solar/pubs/redbook/
Web-based software providing **Photovoltaic Geographical Information System (PVGIS)** for Europe and Africa: http://re.jrc.ec.europa.eu/pvgis/

Heat Pumps

Organisations

North America

Earth Energy Society of Canada: represents the earth energy industry and promotes ground source energy technologies, www.earthenergy.ca/conta.html

Geoexhange Canada: http://www.geoexchange.org/ and Canada http://www.geoexchangebc.com/

Geoheat: provides lists of installers, case studies and information for consumers. Information and services for industry and consumers: http://geoheat.oit.edu/

Geothermal Heat Pump Consortium: a US organisation for the promotion of GSHPs, includes a large number of case studies: www.ghpc.org

International Ground Source Heat Pump Association (IGSHPA): offers a database of accredited GSHP installers in the USA and around the world, www.igshpa.okstate.edu/directory/directory.asp

Europe

British Geological Survey: geological survey enquiries and services, www.bgs.ac.uk/services/services_for_you/homeowners/home.html

European Heat Pump Association: offers a database of European certified heat pump installers, www.ehpa.org

Ground Source Heat Pump Association, UK: www.gshp.org.uk

Heat Pump Association, UK: www.heatpumps.org.uk

International Energy Agency's Heat Pump Centre: offers information, publications and events on heat pumps, www.heatpumpcentre.org/en/Sidor/default.aspx

Photovoltaic Energy

Organisations

North America

American Solar Energy Society (ASESs): www.ases.org

Canadian Solar Industries Association: www.cansia.ca

International PV Equipment Association, US: www.ipvea.org

International Solar Energy Society (ISES): https://www.ises.org/ises.nsf

IPVEA International Photovoltaic Equipment Association (HQ): www.ipvea.org

Solar and Sustainable Energy Society of Canada: www.sesci.ca

Solar Electric Power Association, US: www.solarelectricpower.org

Europe

British Photovoltaic Association: www.bpva.org.uk

European Photovoltaic Industries Association: www.epia.org

Solar Trade Association: has a useful website and a free downloadable Solar Handbook (they cover solar thermal and PV), www.solar-trade.org.uk

Other resources

Estimates for solar (and wind) system costs, generational potential, payback and financial incentives in the US, www.solar-estimate.org

JRC Sunbird PV resource estimation online: http://re.jrc.ec.europa.eu/pvgis/apps4/pvest.php

Online solar calculator estimating potential income for a property from PV FiTs scheme in UK: www.solarguide.co.uk

US Department of Commerce's 'Earth System Research Laboratory' online solar calculators: www.srrb.noaa.gov/highlights/sunrise/gen.html

Wind Power

Organisations

North America

American Wind Energy Association: www.awea.org

Canadian Wind Energy Association: includes an online guide for siting small wind energy systems, www.canwea.ca.

Wind resources from **NASA:** http://worldwind.arc.nasa.gov/java

Europe

British Wind Energy Association: www.bwea.com

European Wind Energy Association: www.ewea.org

Scoraig Wind Electric: information and resources for small scale and DIY wind energy projects, www.scoraigwind.co.uk

Other resources

California's Rebate Program for Wind: www.energy.ca.gov/renewables/emerging_renewables/index.html.

Power Predictor wind and solar energy prediction software, available online: www.bettergeneration.com/power-predictor/introducing-the-power-predictor-anemometer.html

UK Government wind database: www.decc.gov.uk/en/windspeed/default.aspx
US

US DOE: web page with links to information on wind and hydro power, www1.eere.energy.gov/windandhydro/wind_how.html

US DOE: online help with wind assessment by US State, www1.eere.energy.gov/windandhydro/wind_potential.html

US DOE: on measuring potential wind, www1.eere.energy.gov/windandhydro/wind_potential.html

Wind maps USA: www.windpoweringamerica.gov/wind_maps.asp

USA Consumer Guide recommendations and UK fields trials (EST, 2007[1])

The USA small wind turbine Consumer Guide recommends the following key circumstances to consider even before monitoring your wind:

- Is there enough wind resource?
- Are tall towers acceptable or permitted in your locality?
- How much electricity do you need (either for own use and/or to sell to the grid)
- Can you connect to the grid in order to sell it on (or is the grid so far that wind power is a stand-alone alternative)?
- What will the costs of installing and maintaining the turbine system be?

One other point could well be added:

- What are the locally available levels of capital grant incentives, preferential tariff rates for the sale of clean electricity and tax incentives?

Recommendations of the EST Field Study include these guidelines for potential turbine buyers:

- Wind turbines work only when installed properly in an appropriate location;
- They offer potential for delivering carbon savings and energy generation in the UK;
- The highest potential for successful household small-scale wind installations in the UK is in Scotland (because of the strong and consistent winds there);
- Wind speeds are difficult to predict and highly variable;
- Utilise the best available windspeed estimation tools and, if possible, use anemometry to determine the wind speed distribution (for more on this see Basic Physics and Wind Monitoring sections);

1 www.energysavingtrust.org.uk/Generate-your-own-energy/Energy-Saving-Trust-field-trial-of-domestic-wind-turbines

- Buyers are advised to only consider domestic small-scale wind products and installers that are certified under the Microgeneration Certification Scheme (applies to UK only); and
- Householders should consider energy produced from small-scale wind as one option from a potential suite of microgeneration technologies.

The turbines monitored were domestic small-scale building mounted and free standing between 400 W and 6000 W rated output. Some 57 Energy Saving Trust funded installations were monitored plus householders at 68 additional sites provided monthly energy generation data. Additionally, 29 Warwick Wind Trial sites offered data. The data indicated that none of the sites with building-mounted turbines had average annual recorded wind speeds of 4 m/s or greater, and only a third of free-standing pole mounted turbines had average annual recorded wind speeds of 5 m/s or greater. Building-mounted turbines failed to reach the commonly quoted load factors of 10 per cent and no urban or suburban building-mounted turbine generated more than 200 kWh (equivalent to £26 per annum at the time). Some cases showed a net loss of electricity due to the inverter needing power from the mains when not generating.

Free-standing turbine sites had better results, but these were generally in rural locations, often exposed, near coast or on mountains. The results frequently exceeded commonly quoted annual load factors of 17 per cent and the average monitored load factor was 19 per cent (with best sites having load factors of 30 per cent or above). The study suggested that a 6 kW turbine with a 30 per cent load factor would produce around 18,000 kWh per annum (around £2,340). Many free-standing turbine sites had average annual wind speeds greater than the building-mounted situations. Only seven sites had wind speeds greater than 5 m/s. One important conclusion was that free-standing turbines sited in built up areas did not perform as well as those in remote, more windy, locations.

Despite the poor urban performance of turbines, the Energy Saving Trust estimates that for domestic small-scale wind turbine (which they define in the 400 W to 6 kW range) installations there could be as many as 455,650 domestic properties in the UK (which equates to 1.9 per cent of all households) that would have a suitable wind resource of at least 5 m/s and adequate land area and/or building profiles.

Hydropower

Organisations

British Hydropower Association: www.british-hydro.org

Canadian Hydropower Association: www.canhydropower.org

European Small Hydropower Association: www.esha.be

US Hydropower Association: www.hydro.org

Index

absorbers: building materials 25; evacuated tube collectors 78–80, *79–80*; flat-plate collectors 76, *76–8, 77*; heating efficiency 74; pool-heating 73
advice services 16, 63
air conditioning 16
Air Conditioning and Refrigeration Institute 110
air source heat pumps: benefits and disadvantages 106–7, 109; cooling measures *19*, 19; costs and efficiency 108, 202–3, *203*; delivery temperature 109; heating or cooling 96–7, *97*; resource access 7; types and installation 106
American Wind Energy Association 160
Ampair 149
Australia 161, 162, 219
Austria 57, 58

bacterial growth, prevention of 82, 110
battery storage: micro hydro systems 180–1, 191, *191*; photovoltaic output 125, 128; wind turbines 133, 135, 145, 159
Bergey 149, *149*, 151
billing methods 10, 94, 123
boiler installation: capacity and heating needs 45, **46–7**, 61–2; control system 48–9; selection criteria 42–3; wood heating 54–5, 65, 67–8; wood pellets 59, *60*
British Fenestration Rating Council 37
buffer tank 45, 88, 107
building, cooling: air source heat pump *19*, 19; insulation 18; natural ventilation *17*, 17–8; reflective devices 19; shading *18*, 18–9
building, warming: draught-proofing 20–3, *21, 22*; insulation 23–4, *23–4*; solar absorber surfaces 20, 25, *25*–6
business users: equipment efficiency 39; financial support 217; ground source heat pumps 100, 101, *101*; heating management 48–9; hydropower 188–9, 190; lighting efficiency 39; low energy design 193, *194–5*, 195–6;
photovoltaics (PV) 114, *122*, 124, *125*, 127, 196, *196*; Renewable Heat Incentive (RHI) 223; wind power 131, 132, 155

Canada 162, 217–8
capital grants: hydropower 189, **191**; legal requirements 128, **191**; national incentives **8–9**, 215–7; photovoltaics (PV) 128, 209; wind power 147, 159; wood heating 63
Carbon Trust 15, 144
central heating systems: control system 48–9; wet-heat distribution 100, 106, *201*, 201–2
Centre for Alternative Technology: hydropower 168, *172*; insulation 20, 30; solar water heating 79
CO_2 emissions: Carbon Emissions Reduction Target 217; heat pumps 94, 107; payback periods **222**; photovoltaics 6; renewables company *197*; UK energy consumption 15; wood pellets 57
collectors: clip-fin *76*, 76, 78; efficiency and siting 74–6; evacuated tube 78–80, *79–80*, 89; flat-plate 76–8, *77, 78*, 89, 199
community and shared renewables 11; financial support 216–7; heating systems 69; hydropower 189–90, 212–4, *213*; wind turbines 154–5, 211, *212*
conservatories/sun room 26
cube law 137

Department of Trade and industry (UK) 2, 57
desktop survey 176
differential thermostats 84
district heating networks 11, 69, 88
Domestic Heating-Design Guide 44
doors: frames and draughts 21–2; insulation 37
draught-proofing 20–3, *21, 22*
Dulas Ltd 193, *193–7, 195–6*

electronic load controllers 173, 175
emission regulation: national guidelines

62–3; *see also* CO_2 emissions; wood heating installations 66–7
energy, definition of 10
energy efficiency: cooling measures 16–9, *17–9*; domestic appliances 40–1; energy monitoring 37–8, *38*; global measures 14–5; heat distribution systems 44–5, **46–7**, 47–9; heat pumps 91, *92*, 93–4, *94*; humidity and temperature control 41–4; insulation materials 27–8, *28*, **29**, *30*–2, **32**, 33–5, *34*–5; lighting 38–9, *39*; meaning of 13–4; office equipment 39; passive solar design 24–7, *25*–6; refrigeration 40; stand-by loads 41; warming measures *20*–4, 20–4; windows and doors *35*–6, 35–7; wood stoves and boilers 66
Energy Efficiency Partnership for Homes 42
energy monitoring 37, 37–8, *38*
Energy Policy Act (1992) 161, 189
Energy Saving Trust: advice services 15; heat pumps systems 95, *95*; insulation 24, 32; payback calculation 221; photovoltaics (PV) deals 217; wind turbines 144, 146, 147
Energy Star Program 15, 109
Environment Agency 66, 190
Europe: emission regulation 62; energy efficiency 15; installation regulations 68; solar radiation *72*; wood pellet standards 66
Evance 151, *152*
expansion vessels 85

feed in tariffs (FiTs): national incentives 8, **8–9**; payback calculation 220–2, **222**; payback schemes 218–20; photovoltaics (PV) 111, 115, 124, 127, 129–30; photovoltaics (PV), domestic *120*, 202, 206–7, 210; wind power 131, 135, 150, 155, 158, 161
financial support: national sources of 215–25; *see also* capital grants; feed in tariffs (FiTs); tax credits
flagging (tree) *141–2*, 141–2
floors: insulation 34, *34*–5; underfloor heating 48, 89–90, 94, 98, 100, 109
fuel, availability and price: comparison of 64, **70**; wood 60; wood pellets 58
fuel-feed mechanisms 60, *61*

geothermal heat pumps *see* ground source heat pumps
Germany 130
Green Deal for Homes 223

Griggs-Putnum Index of Deformity 142
gross metering 124
ground source heat pumps: benefits of 101; equipment, selection criteria **102–3**; heat distribution 98, *99*, 100; heat loop installation 98, *99*, 100, **103**; installation, reasons for 203–5, *204*; site suitability 7, *8*, 102, 205

heat exchanger 81–2, 93
heating systems: boiler choice and operation 42–4; distribution systems 44–5, **47**, 47–9; heat-load calculations 45, **46–7**; heat-loss calculations 43–4
heat pumps: air source 105, 106–8; Coefficient of Performance 91, *92*, 93–4, *94*; common types of 95–7, *97*; equipment, selection criteria 108–10; function and operation of 91–3, *92*, *93*; ground source 98, *99*, 100–3; water source *103*–4, 104–5
HETAS 67
Hockerton Housing Project 211, *212*
home appliances, efficiency of 40–1
hydropower: advantages of 188; catchment and flow 169–71, *170*; community projects 189–90, 212–4, *213*; debris removal 183, *184*, 186; flow calculation and power output 177–80, *179*, *180*; grid-connected *175*; history of development 165–8, *167*; hydrological cycle *164*, 165; hydro scales 164; impulse turbines 172–3, *173*; intake and powerhouse, siting of 176–7; intake and water diversion 182–3, *182–3*, *185*; issues to consider **191**; limited availability 3; micro systems 180–1, *180–1*, 186; off-grid system **191**, *191*; penstock design and installation 184–6, *185*; powerhouse 187, *187*–8; regulations 190; site suitability 168–70, *169*, *170*, 176, *177*; system components *168*; turbine efficiency 171–2, *172*

insulation: choosing materials 28; cooling measures 18; ground floors 34, *34*–5; lofts and roofs 34; R-values 24, 27; wall techniques 32, *33*; warming measures *23*–4, 23–4, 200; windows *35*–6, 35–7
insulation materials: application and effectiveness **32**; natural fibre *28*, **29**, *30*; product forms 31, *31*; rigid foam 31
intake and water diversion 182–3, *182–3*, *184*

inverters: grid-connected (hydro) 181; grid-connected (solar) 122, *123*; grid-connected (wind) 135; off-grid system 125, *126*; operation of 113, 129; shading problems 128

J Manual Residential Load Calculation 44

lighting 38–9, *39*
lofts, insulation of 34
Low Flow 2000 178

Marlec *136*, 149, *150*
Microgeneration Certification Scheme 110, 128, 161, 218

National Energy Policy Act (1992) 161, 216
National Fenestration Rating Council (US) 37
neighbours 131–2, 159, 188
net metering: Australia 161; hydropower 189; national incentives 8, **9**; photovoltaics (PV) 115, 124, 127, 207; United States 129, 157, 189, 207; wind power 150, 157
new build home: low energy design 197, *198*, 199; wood pellet heating 199
noise 107, 159
North American Board of Certified Energy Practitioners (NABCEP) 128
Numerical Objective Analysis Boundary Layer (NOABL) 143–4

office equipment, efficiency of 39

Passive House Planning Package (PHPP) 26
passive solar design 24–5, *195*, 195–6
passive solar systems 85–6, *86*
payback periods: calculation method 220–2, **221**; photovoltaics (PV) 206–7; solar water heating 73–4, 87; wind power 159, 161
payback schemes: national incentives **8–9**, 215–7; *see also* feed in tariffs (FiTs); net metering
Pellet Fuels Institute 66
phase changing materials (PCM) 25
photovoltaics (PV): advantages of 114; business users 124, *125*, 127, *196*, 196, 207–9, *208–9*; construction and operation 112, *112–4*, 116, 116, 208–9; costs and financial support 128–30, 202, 206–7, 209, *210*; energy output 115, *209*, *210*; equipment, selection criteria 126–8; global use of 111–2; grid-connected house *112*, 122–4, *123*, 210, *210*; installation contractors and regulation 127–8; markets for electricity 115; module mounting systems 120, *120–2*; module types and efficiencies 116–9, *117–9*; off-grid system 125, *126*; site suitability 6, *6*
power, definition of 10
Proven 132–3, *136*, **153**, *153*
Public Utilities Regulatory Policies Act (1978) 135, 161, 190, 216

radiators: delivery temperature **109**; low-temperature **102**, 106, 108; outputs 43; wet-heat distribution 47, 48, 89
Renewable Heat Incentive (RHI) **8–9**, 222–4
renewable resources: costs and financial support 8, **8–9**; independent generation 10–1; technology and site suitability 3, **4**, 5–7
roofs, insulation of 24, 34
R-values: calculation method 26–7; insulation materials 24, 27

Scandinavia 54
shading: cooling measures *18*, 18–9; passive solar design 25
site suitability: ground source heat pumps 7, *8*, 102, **103**, 205; hydropower 168–70, *170*, 176, *177*; photovoltaics (PV) 6, *6*; renewable resources **4**; solar thermal installations 6; wind power 131–2, 137–9, *139–40*, 143
solar liquid heating 89, 89–90
solar radiation 71–2, *71–2*
solar space heating 87–9, *88*
solar thermal installations: pool-heating 73; site suitability 6, *6*; water heating 45
solar water heating: collector/cylinder *81*; collectors, efficiency and siting 74–6; control system 84–5; costs and financial support 202; direct system *83*, 83; drain back system 85; dual coil tank 81–2, *82*; economics of 87; equipment, selection criteria 86–7; evacuated tube collectors 78–80, *79–80*; flat-plate collectors 76, 76–8, *77*, *78*; fully filled thermal system 84, *84*; heating process 75; integral collector storage systems 80; multi-residential building 200; Renewable Heat Incentive (RHI)

222–4; thermo-syphon system 85–6, *86*
Southwest Windpower *136*, *149*, 149, *151*
stagnation, causes of 87
stoves: masonry 55, *56*; wood log *52*, 54–5, 65, 67, *67*
sun pipes 25
suppliers and installers: air source heat pumps 109; boiler installation 44; certification 219, 223; heat pumps 96, 101, 110; photovoltaics (PV) 127–8; solar water heating 86; wind power 143–4, 159, **160**; wood heating 55, 62, 63, 67–8

Talybont Energy 212–4, *213*
tax credits (US) 190, 207, 216
thermal conductivity (K-value) 26–7
thermal resistively *see* R–values
thermal transmittance *see* U–values
ThermaSkirt 48
Thermomax 78–9
Trombe wall *25*, 25
turbine, water powered: electronic load controllers 173, *175*; head calculation and output 176–7; impulse 172–3, *174*; pelton wheel 173, *173*, 187, *187*; reaction 172, 173, *174*

underfloor heating: air source heat pumps 98; delivery temperature **109**; fitting of 34; ground source heat pumps 100; solar liquid heating 89–90; wet-heat distribution 48, 94, *199*
United Kingdom: domestic consumption 2, 16, *16*; domestic electricity generation 124; emission regulation 66–7; energy efficiency 15, 37; financial support 216–7; hydropower incentives 188; installation regulations 55, 67–8, 128, 162, 190; photovoltaics, support for 129, 130; wind farms 155–6, *156*; wind power incentives 158, 161; wind speed data 143–4
United States: domestic consumption 2, 16, *16*; emissions regulation 62, 66; energy efficiency 14–5, 37, 66; financial support 216; home energy uses *16*; hydropower incentives 189; installation regulations 67, 68, 128, 162, 190; net metering 124, 129, 157, 189, 207; photovoltaics, support for 130; renewable energy schemes 1–2; solar radiation *71*; tax credits 190, 207, 216; Tribal Energy Programme 216; utility regulation 135; wind power guidelines 147; wind power incentives 157, 161; wind speed data *142*, 143; wood pellet standards 66
US Department of Energy: air conditioning costs 16; Energy Efficiency & Renewable Energy initiative 14–5, 216; insulation materials 28–9; solar water heating 81; water heating usage 74; wind speed data 143; wind turbine output 151
US Environmental Protection Agency (EPA) 66
U-values, calculation method 26–7

valves, safety: heating management 49; solar water heating 85; water powered turbine 185
ventilation: heat recovery 42; natural *17*, 17–8
voltage optimization 37–8, **221**

walls, insulation techniques *32*, 32
water source heat pumps *103–4*, 104–5
windows: double or triple glazing 36–7; external storm *36*; frames and draughts 22; passive solar design 25
wind power: community projects 211, *212*; feasibility and installation **160**; grid-connected building *135*, 135; grid-connection and incentives 158, 160–1; heating systems 135–6; issues to consider 131–2, 147, 159; output calculation 137; planning regulations 162; stand-alone systems 135; turbine costs and selection **158**, 158–60; turbine design 132–5, *133*, *134*; turbine location and selection *5*, 5, 138–9, *139*, *140*, 142–3, 157; wind resource assessment 140–4, *143–4*, 148; wind speed and effect **138**, *141–2*
wind turbines: cut-in speeds 142; horizontal-axis 132–4, *133*; large-scale (wind farm) 155–6, *155–6*; medium-scale 154–5, *155*; micro generators 148–9; operation of *136*, 136–7, 145; output terminology 145; power curves 146, *146*; small generators 150–1, *151*; towers and cabling 156–7; vertical-axis 134–5, *135*
wood heating: disadvantages of 68; energy efficiency 66; equipment, selection criteria 61–3, 70; fuel costs 60, *64*, 64, **70**; human input and demand 65; log stores *53*, 54; log stoves and boilers 54–5, 65, 67, *67*;

modern systems 51–2, *52*, 54; resource access 6–7
wood pellets: emissions 57; heating systems 56–7, 60, *60*, *61*, 199–202, *201*; manufacture 56, 57; national standards 66; storage 58–9, *63*, 199; stove design 52
World Energy Council 6